Economy and Environment The Enigma of Health and Prosperity

I0473662

Jon Frank Anderson

CHAPTER 1: ESSENTIAL CLEAN AIR ACT INFORMATION

CHAPTER 2: ENVIRONMENTAL ACTION AWAY FROM THE HEADLINES

CHAPTER 7: THE UNITED NATIONS FRAMEWORK ON CLIMATE CHANGE...136

CHAPTER 8: RENEWABLE VERSUS CLEAN ENERGY.156

CHAPTER 9: CLIMATE CHANGE CONTROL STRATEGIES 163

Chapter 1: Essential Clean Air Act Information

What are the National Ambient Air Quality Standards (NAAQS)?

What are the National Ambient Air Quality Standards (NAAQS)? Air pollution control programs in the United States depend on the attainment status of the metropolitan area. Areas that are in attainment of the National Ambient Air Quality Standards (NAAQS) do not require air quality control programs. Areas not meeting the NAAQS are non-attainment areas and must adopt control programs depending on the regions severity of pollution.

The NAAQS are set for "Criteria Pollutants" of Ground Level Ozone (O_2), Carbon monoxide (CO), Nitrogen Oxides (NOx), Sulfur Dioxide (SO_2), Lead (Pb), and Particulate Matter smaller than 10.0 microns and 2.5 microns. There are primary standards (health based) and secondary standards (property damage based). Standards are measured over averaging times in concentrations expressed as parts per million (ppm) or microns (ug/m3).

Under Section 109 of the Clean Air Act, EPA must set standards to protect public health. This includes sensitive populations such as children, the elderly, outdoor workers, asthmatics and people with cardio-pulmonary heart disease.

The standard for Ground Level Ozone was tightened from 0.12 ppm to 0.08 ppm for a maximum one hour average. The Bush Administration relaxed the standard back to 0.12 ppm. Obama re-adopted the new, more protective standard of 0.08 ppm.

Particulate Matter has long had a bi-modal distribution with the traditional peak around 10 microns (micro meters) and another around 2.5 microns. PM10 is filtered naturally by the body, for instance sand. PM2.5 is fine particulate that air toxics like toluene and benzene adhere to causing lung cancer pre-maturely. Ammonia and formaldehyde (HCHO) from cattle urea also adhere to PM2.5 causing and exacerbating Acidic Precipitation of Nitrogen Oxides (NOx) and Sulfur Oxides (SOx).

Since Lead was removed from gasoline in the 1980's Lead NAAQS violations are rare. Carbon Monoxide is also much less of a problem as most areas are now in a Maintenance status of the Carbon Monoxide (CO) NAAQS thanks to the success of the Federal Tier 1 and Tier 2 tailpipe emission program and other Clean Air Act of 1990 (CAA 90) mobile source control programs.

Even Acidic Precipitation is decreasing thanks to the success of the Sulfur Oxides (SOx) emissions trading rule and the newer Clean Air Interstate Rule (CAIR) allowing the trading of Nitrogen Oxides (NOx).

Air toxics and Greenhouse Gases (GHGs) are not original Criteria Pollutants and do not have NAAQS set for them. GHGs include Carbon Dioxide (CO_2), Methane (CH_6) and Nitrogen Dioxide (N_2O).

There are three other chemicals, emitted in smaller amounts and with shorter half lives than Carbon Dioxide's several hundred years' half life. The half life of Methane is measured in several months, not hundreds of years.

What Are Volatile Organic Compounds (VOC) and Air Toxics?

Volatile Organic Compounds or VOCs are a class of about 120 Hydrocarbon chemicals. Air toxics are carcinogens. Petroleum is mostly VOC with a great amount of toxics.

Air toxics are found every where every day in the air we breath and the water we drink. Air toxic from petroleum are also absorbed through the skin.

Toluene and benzene are the most commonly mentioned toxic. Toluene is present in all solvents and anything with petroleum in it. There are other groups of toxics in VOC that received press coverage during the Deepwater Oil Drilling Platform catastrophe in the spring and summer of 2010. The dispersants used to keep the oil spill in the ocean and not on shore released Poly-Aromatic Hydrocarbons or PAH, which are highly cancerous.

I am not a scientist, chemist or botanist. Neither am I a fisheries or wildlife expert. My background is transportation air quality, not water. I am not an expert in oil rigs or utilities or refineries. I am not an expert in meteorology or climate science. I am an analyst or consultant on a contract basis.

Scientists of all stripes were tainted by social media and social change. Just by change. Recently scientists have seized upon a chance to act and make a difference.

Political advocacy is not the scientist' job, research and reporting in a neutral manner is the scientist job. Their job is objective science and reporting of facts.

The lobbyists, commentators on TV, consultants and the check writers who follow government at any level are the ones who interpret science from reports to convey to elected and appointed public officials.

I have used the same research on an issue for the same political party in different states and have seen opposite action taken in the various states. I have witnessed Republicans support a measure in one state and Democrats oppose the same measure in another state. All politics is local, the Great Tip O'Neil said

VOC evaporates. Too quickly and it explodes, too slowly and it pollutes. VOC from power plants and tail pipes in the summer mixes with Nitrogen Oxides in the air to create Ground Level Ozone. This happens in hot temperatures in the summer in Temperate regions of the world and all the time in the Tropics. Ground Level Ozone pollution is exacerbated during an air inversion or stagnation when temperature differentials at varying parts of the lower atmosphere lock the pollution in place for days and sometimes weeks.

VOC is easily controlled to reduce ground level ozone problems. VOC control is a modest success of the Clean Air Act.

Air Toxics are impossible for me to address because they are too vast, complex and less political than other air pollution problems. Air pollution control rules on cars, fuels and power plants are always highly charged and political, often garnering press scrutiny. Air toxics are nasty, bad stuff innate to hydrocarbons from petroleum.

But petroleum is the prime mover of our economic, mobility, security and safety. The social nuisance and external environmental cost of petroleum on health and property damage is not a welcome conversation topic. Ground Level Ozone and Particulate Matter or soot dust, cause asthma, cancer and worsens people with other existing heart and lung disease.

But unlike Carbon, VOC is mostly controlled now. You pay for VOC control cost every second and time you flip a switch, board a transport vehicle or start an engine. Most VOC control was achieved through great progress in business process improvement and technology. VOC control was done several times cheaper than the United States Environmental Protection Agency (U.S. EPA) and industry originally estimated in the early Clean Air Act years.

There are no known technologies to control Carbon Dioxide. Some say Emissions Trading is a paper transaction charade on Wall Street like derivatives and mortgages. What are some of the Carbon Control ideas often simple and old but labeled with new names?

Offsets are a commonly mentioned Carbon Control strategy.

Offsets in the minds of some are forestry and agricultural management practices not in use and increased. A good example is re-forestation in the Tropical Rain Forest.

Energy Conservation and Energy Efficiency is a return to "as built" operation. This maintenance is not always done. It does decrease Carbon pollution. Both of these frequently mentioned Carbon Control strategies do decrease Carbon levels.

There are not many Carbon Controls to undisputedly reduce Carbon. Other skeptics to Carbon Control have little faith in government laws and rules to control Carbon. Almost everything has Carbon in it. There will always be Carbon in the air.

Can we control Carbon? Most of this book discusses Carbon Control from differing points of view as unfolded on the national and international stage in 2009 and 2010.

Should we give up on Carbon Control efforts? No. It gives us purpose and something to fight about and maybe really solve like Upper Level Ozone that scared me in my nightmares as a child, and now, as a close friend tells me, her young child fears Global Warming.

Is that what this is about fear? Is Carbon Control about denial or solutions, on paper, in Tropical Rain Forests, in Energy audits, in Carbon burying and storage and in non-fossil fuels? We hope our efforts and unknown results in centuries from now is worth the time and cost now.

The Clean Air Act as a Success

Environmental protection is perhaps the biggest government domestic success of my generation. We rarely stop to think of pollution. The Clean Air Act, first in 1970, again in 1977 and then permanently in 1990, has six Criteria Pollutants, and now, according to the Supreme Court, six Greenhouse Gases.

The original six Criteria Pollutants are Carbon Monoxide (CO), Sulfur Dioxide (SO_2), Nitrogen Oxides (NOx), Volatile Organic Compounds (VOC, about 120 compounds), Particulate Matter (PM10 and PM2.5) and Lead (Pb). The new greenhouse gases are Carbon Dioxide (CO_2), Methane (CH_4), Nitrous Dioxide (N_2O), Hydrofluorocarbons (HFCs), Perchlorofluorochlorine (PFC) and Sulfur Hexafluoride (SHF). The primary focus is on CO_2 and Methane. All greenhouse gases (GHG) are called Carbon Dioxide equivalents (CO2e).

The original six Criteria ambient (outdoor) air pollutants are reasonably controlled since 1970, most notably Lead and Carbon Monoxide. Lead was phased out of gasoline in 1980's and is now a rare, local stationary plant issue.

Carbon Monoxide is not as big of problem thanks to the transportation and vehicle controls from 1990. Volatile Organic Compounds and Nitrogen Oxides are also better controlled. Sulfur Dioxide was solved by the original Cap and Trade program that came in far below industry and EPA estimates $1,000 to $1,500 per ton (2,000 pounds). Sulfur now trades for around $700 per ton.

The Northeast Regional Greenhouse Gas Initiative (RGGI) is a voluntary state Carbon Cap and Trade program selling Carbon for about $3 per ton to industry. The West and Midwest are starting similar programs. Carbon should remain around $15 per ton for several years.

Signed by President George H. W. Bush on November 15, 1990, the Clean Air Act, as Amended in 1990 was the first major environmental protection law passed by Congress and signed by the President with no "sun-set" or expiration date, like every other major environmental protection act between 1970 and now.

As we go forward with Carbon Control programs, functioning in half of the United States through voluntary state compacts, air quality control planners should not forget Ground Level Ozone and Particulate Matter or soot dust from fuels. Many Carbon friendly fuels and options in study, evaluation, deployment and assessment are not heart-lung friendly options despite lower Carbon content.

Carbon Control must not sacrifice the more mundane problem of summer Ground Level Ozone weeks for the majority of the United States population which breathes and walks laboriously outside on Ozone pollution days.

In planning for Carbon Control options air quality control planners need keep in mind the Ground Level Ozone suffers of asthma, the elderly, outdoor workers, children, cancer survivors and those with heart or lung impairment. A majority of the nation's citizens need protection from the old, common and known air pollution problems.

EPA Tightens Ground Level Ozone Standard

EPA tightened the Ground Level Ozone standard in 2009. The EPA enacted rules tightening the National Ambient Air Quality Standards (NAAQS) for Ground Level Ozone pre-cursors, Volatile Organic Compounds (composed over 120 chemicals) and Nitrogen Oxides. Ground Level Ozone is the most widespread air pollution problem in the U.S. affecting two thirds of the population.

This tightening of the standard is rare, only the third time in the 40 year history of the Clean Air Act. The rule will cost industry billions but prevent thousands of premature deaths annually sensitive people.

The new 8 hour standard is 60 parts per billion (ppb), lowered from 75 ppb. There are 1 hour and 8 hour averaging times for the standard. The rule poses a big air quality challenge for state and local governments to attain.

The rule will be phased in over 20 years depending on the metropolitan region's air quality non-attainment status. Severely polluted cities like Philadelphia, New York City, Los Angeles, Chicago, Houston and Dallas will have the most time to comply. Boston and the rest of the Northeast will have to comply sooner, but with less stringent programs. Look for mandatory car pooling to raise its ugly, political head again in the next two decades.

The American Petroleum Institute, the oil industry lobby, said the rule, "lack scientific justification".

EPA's Clean Air Scientific Advisory Committee recommended unanimously the standard tightening in 2007. President Bush loosened the standard, ignoring the panel, in 2008, citing the economic impact and job loss.

The American Lung Association sued over the Bush Ground Level Ozone standard rules. EPA consented in September 2009 to review the Bush rule. The standard was set in 1971 under the Clean Air Act, loosened in 1979, and tightened in 1997.

The rule will cost $19 to $90 billion by 2020 according to EPA. The estimated economic benefit is valued between $13 and $100 billion. The rule will save more than 12,000 people from premature death by 2020.

Research has proven children who grew up in polluted areas where Ground Level Ozone concentrations are high had severe physical developmentally impaired lung capacity compared to children in less polluted areas. This made them more vulnerable to early, life long asthma, respiratory illness and premature death.

Fine Particulate Matter

Fine Particulate Matter was found deadlier than thought. The Health Effects Institute of Cambridge, MA released a study showing the impacts of fine Particulate Matter smaller than 2.5 microns is deadlier than thought and requires more regulation. The EPA has been regulating PM2.5 since 1997.
The Bush Administration declined to review new studies. The current EPA under Administrator Lisa Jackson is examining whether a new PM standard is needed. Many studies show the threshold should be set at around 20 nanometers, a bout a 30th the size of a human hair.

In 116 cities around the country, heart attack risks increased from 12 percent to 24 percent in areas with elevated fine PM. The analysis covers 350,000 people over 18 years, with another 150,000 added in recent years. The study was undertaken by the University of Ottawa for the Health Effects Institute. Fine PM or soot comes from coal power plants, diesel and gasoline, car tires and oil refineries.

PM penetrates the lungs deeply and adheres to fine air toxics or carcinogens bringing them into the body. PM also causes asthma in both urban and suburban areas. It elevates the risk of pre-mature death from those suffering cardiopulmonary heart disease and lung problems.

Particulate Matter is a problem for the young, the elderly, outdoor workers, people with asthma, and people with heart and lung disease. The government values the life of a person at $6 million. The cost of preventing pre-mature mortality is cost effective and has prevented hundreds of thousands of pre-mature fatalities.

Corn and Ethanol Subsidy Saga

Corn based ethanol subsidies could end. EPA, under its Climate Change rulemaking initiative is proposing to end the tax subsidy to Midwestern corn producers. However, only the Congress can end the subsidies, and that is unlikely to happen. EPA is proposing the indirect emissions from bioethanol be included in the overall emissions calculation. Full life cycle emissions calculations have been in use for two decades by analysts, but are only now adopted by EPA. EPA predicted that as much as 10 to 15 percent of increased food costs in the last several years can be linked to corn based ethanol production.

As expected, Midwestern corn growers were upset by the ruling while the Cattlemen's Feed Association was pleased by the ruling. The price of corn based feedstock for livestock has risen considerably since the government urged corn based ethanol for transportation fuel 10 years ago.

The indirect costs of corn based ethanol include clear cutting and deforestation in the Emerging World as the farmers scramble to gain a foothold in the market. This indirect cost outweighs the benefits of the lower Greenhouse Gas emissions from corn based ethanol. The inclusion of indirect land use emissions from machinery used to grow and transport fuels such as diesel and gasoline make it impossible for corn and soybeans to pass the full life cycle emissions test. Other bioethanol fuels such as switch grass, algae, and even garbage easily pass the full life cycle emissions test.

California recently passed the first in world regulation to reduce the amount of Carbon Dioxide released by transportation fuels. The goal is to cut Carbon emissions from transportation fuels by 10 percent by 2020 and by 80 percent by 2050. The $400 billion annual transportation Energy fuel cost goes mostly overseas in the import of petroleum.

The move to reduce the Carbon footprint is widely seen as a move that would improve the U.S. economy, because low Carbon fuels are easily produced domestically, and would foster adoption of new technology and create thousands of jobs.

In a similar move the Obama Administration and EPA then considered and adopted a similar nationwide standard to reduce the imprint of transportation fuel.

Even some large oil companies are on board with the Carbon reduction plan, seeing it as way to diversify their investments while remaining competitive with emerging companies that could capture substantial market share from the petroleum giants. The California rule goes into affect in 2011. However, do not look for the Congress to cut corn subsidies to the Midwest any time soon. Attempts have been made in the past to end the corn subsidy and they all have failed.

Upper Level Ozone Hole Healing

The Upper Level Ozone hole is improving. Monitoring of the world's Upper Level Ozone hole created by Chlorofluorocarbons (CFC) is closing. This is viewed as good news by government officials around the world that banned production and use of CFCs in 1987 in the Montreal Protocol. There appears a scientific dispute about the improvement of the Upper Level Ozone layer on Climate Change.

The hole in the Ozone Layer created thicker than normal cloud cover in the Antarctic regions, blunting the effect of temperature change. A Geophysical Letters report noted some scientists said the reduced cloud cover will increase the effective of temperature change. Apparently, according the U.S. National Oceanic and Atmospheric Administration, Upper Level Ozone levels affect the earth's wind speed. Higher wind speed picks up more salt from the oceans leading to thicker cloud cover.

Scientists who wrote the paper say the Upper Ozone Level recovery will reduce wind speed and cloud cover, increasing the effect of temperature change. U.S. government scientists say increased Carbon Levels in the atmosphere has the same effect on wind speed and cloud cover. So far, the evidence on each side is incomplete.

Ozone in the upper level of the atmosphere protects the earth from ultraviolet radiation from the sun, affecting skin cancer rates.
Ground Level Ozone is not protective and a dangerous irritant to the lungs of the young, elderly, outdoor workers, athletes, people with asthma, heart, lung and cardiopulmonary disease and diabetics.

Ground Level Ozone has decreased due to the success of the Clean Air Act. Cancer rates are down by two percent annually in the U.S. since 1990.

It is about 20 years since governments around the world agreed on the last major world wide environmental treaty. The Montreal Treaty banned Ozone depleting chemicals in the upper atmosphere. It is the last time domestic national governments of all ideologies outlawed a harmful chemical at the local level.

It is a resounding environmental policy success that has dramatically eased skin cancer.

A friend told me in 1993 the replacement coolant for his car's air conditioning was not as cold as the old the refrigerant was.

There is product functionality loss and quality degradation with cleaner, alternative, non-deleterious chemicals. The replacement chemicals reduce cancer causing, organ specific and systemic immune-suppressant health receptors in living animals and people.

The product efficiency discount is coupled with an economic cost to the maker who increases the consumer purchase price. This is the trade off for public health and private property protection from toxic natural and man-made chemicals.

But the economic Willingness to Pay Theory is an old postulate of the school of internalizing the external environmental societal costs from pollution. In poll after poll, the American public and people in the Developed World, support and confirm the Willingness to Pay Theory. The public is willing to pay a few more pennies and suffer a marginal product utility discount for improved public health, private property protection and longer life.

I knew 20 years ago when I left certainty and job security of civil service for private consultancy that my chance to help plan and implement the Clean Air Act as amended in 1990 was a once in a lifetime risk worth taking. I hoped to work for cleaner air, reduced cancer morbidity and mortality and improved heart-lung function for the average person.

Sometimes it is better to have tried and failed than never try at all.

The Upper Level Ozone hole is healed, decades ahead of all NASA models predictions in 1990. Coordinated and implemented world environmental protection is cost effective, creating new solutions and jobs and protecting public health. The vanishing of a global skin cancer pandemic is benchmarked and proven.

Cash for Clunkers

How did the Cash for Clunker Program in the summer of 2009 affect displacement of gasoline and Carbon? When hearing of the success of the Cash for Clunkers Program, curiosity prompted quantification of the impact. This commentary is primary technical analysis.

Assuming $2 billion more in is injected to the Clunker Program, removing 750,000 gas guzzlers from the road with a 2/3rds of them getting 4 mile per gallon (mpg) increase and a 1/3rd of them the 10 mpg fuel economy increase, with the average U.S. passenger pegged at an average of 25 mpg, driven 60 miles per day on average across the country, or very roughly 20,000 miles per year with Carbon emissions at 350 grams per mile.

At 25 mpg U.S. average fleet wide fuel economy, a car consumes 800 gallons of gasoline per year. An 18 mpg car consumes 1,111 gallons of gasoline per year. With the minimum Clunker 4 mpg improvement to 22 mpg, 909 gallons of gasoline are used, or savings of 202 gallons of gasoline per car per year. At 28 mpg in the 10 mpg increase for the Clunker Program with the $4,500 rebate, 714 gallons of gasoline are used in a year or a savings of 397 gallons of gasoline per year.

Assuming my 2/3rd 4 mpg improvement and 1/3rd 10 mpg improvement, each portion roughly saves 100 million gallons of gasoline per year for a total of 200 million gallons of gasoline per year displaced through the Clunker Program with $3 billion invested, and 750,000 gas guzzlers off the road and permanently scrapped. According to the U.S. Department of Energy, Energy Information Agency (http://www.eia.doe.gov/), the U.S. transportation sector for cars uses 51,092,000 gallons of gasoline per day. The Clunker Program will reduce this amount by 548,000 gallons per day or offset gasoline use by 1 percent.

In terms of Carbon equivalent reductions from the Clunker Program, I assume the Carbon content of one gallon of gasoline driven one mile is approximately 350 grams per mile. If the average driver puts 60 miles per day on their car and get the average U.S. 25 mpg, they emit roughly 840 grams of Carbon Dioxide per day. An 18 mpg car creates 1,167 grams of Carbon per day. A 22 mpg car emits 952 grams of Carbon per day and a 28 mpg car emits 750 grams of Carbon per day. The Carbon savings for a new 22 mpg car is 215 grams per day; a 28 mpg car saves 417 grams of Carbon per day. Doing the math and trying to show the work, the 500,000 new 22 mpg cars save 10.4 kilograms per year in Carbon or 23 pounds of Carbon per year. On a daily basis this is 0.028 kilograms per day or 0.064 pounds per day of Carbon reduced in total. The 250,000 10 mpg cars roughly save the same amount of Carbon, 10.75 kilograms per year or 24 pounds of Carbon per year.

So, what does this mean to total U.S. Carbon and total world wide Carbon emissions output and reduction from the Cash for Clunkers Program? According to the latest available data from the U.S. EPA Climate Change Center (http://www.epa.gov/climatechange/index.html) in 2007 the total U.S. output of Carbon equivalent Greenhouse Gases (including Methane, Carbon Hexafluoride, Nitrous Dioxide and other smaller Greenhouse Gases) was 7.159.1 Teragrams (Tg) Carbon Dioxide (CO_2) Equivalent (Eq) or 7.2 $TgCO_2Eq$. The $TgCO_2Eq$ is called by the U.S. Department of Energy a Million Metric Ton of CO_2 or $MMTCO_2$. Transportation accounted for 1.887.4 $TgCO_2Eq$ in 2005. We will come back to this after we look at the world wide levels of CO_2Eq.

According to the Energy Information Agency in 2005 the developed world referred to as the Organization for Economic Development and Cooperation or OECD, which includes North America, OECD Europe, Japan and Australia/New Zealand, the OECD Group accounted for 13.565 Tg CO_2Eq in 2005 or 48 percent of world wide CO_2Eq emissions. The remaining 52 percent came from the Developing World.

The Cash for Clunkers Program will reduce Carbon by 21.5 kilograms per year. In a world wide Carbon perspective, let's look at in Teragrams. A Tergram is 1 trillion grams. So, the Clunker Program reduces Carbon equivalent Greenhouse Gas emissions by, take a breath, 0.000.000.021 Teragrams per year or 0.000.000.000.058 Tg per day. It's a small world, with much progress needed.

25

Cash for gas guzzlers was also included in the House Energy bill. The Energy bill would give cash to owners of older, gas guzzling cars and trucks if they purchase a new, more fuel efficient vehicle. Legislators hailed the plan. However, the broader issue of the economic impact of a market based Carbon Cap and Trade program have proved thorny for Congress to resolve. Obama wants an Energy bill passed and gave Congressional leaders wide latitude in crafting a bill that will pass in the House and Senate.

As expected, Congressmen in the industrial heartland, petroleum and coal producing states were reluctant to pass a Climate Change bill, fearing it would hurt their home state economies. The hold out states reduced the Carbon reduction target and chose to give some or all of the Carbon allowances away for free. Environmental groups were against giving away the permits and wanted the fees raised to go to Clean Energy research. Some Democrats wanted the 2020 target cut from a 20 percent reduction to six percent. Again environmentalists opposed such a compromise.

The gulf between the two sides was deep. Health care reform took over the headlines in the daily news. Little mention was made of the Energy bill or Climate Change.

The environment and Climate Change are not priorities for the general public, even though awareness of Climate Change and the general agreement of the government need to take action has increased since 1997 when the Kyoto Protocol was signed.

The Senate could not vote to approve a Climate Change bill due to the Great Recession.

Congressmen are aware that the Supreme Court gave EPA the authority to regulate Greenhouse Gas emissions in the lawsuit Massachusetts versus EPA in July 2007. The Court said that Greenhouse Gases are Criteria Pollutants under the Clean Air Act and gave EPA broad authority to regulate them. So Congress did not pass a Climate Change bill and left it up to EPA and the Courts to work out the details.

Congressional action is preferred because it would directly instruct EPA how to proceed and give the Courts clearer legal rational in deciding the inevitable lawsuits that will be filed against Climate Change rules.

Are we sacrificing the environment for an improved economy? Congress passed the Cash for Clunkers bill under steep opposition from environmentalists. This was largely seen as a victory for the auto industry. At the heart of the matter was how to tackle long neglected environmental needs while jump starting the economy. The Obama Administration and the Democratic Congress walked a tight rope between environmentalists and powerful American industries that suffered in the Great Recession. Industry was the winner, as long neglected environmental goals espoused by Obama in the campaign have fallen by the way side as the environment takes a back seat to the economy.

The cash for clunkers program was scaled back from $4.5 billion to $1 billion. The program was in effect from July 1, 2009 through November 1, 2009. Gas guzzlers had to receive less than 18 miles per gallon, but some light-duty trucks qualified even if the fuel economy improvement was only one mpg.

Car and truck models needed to be new, not used. The vehicle owner must have had title to the vehicle for at least a year. Owners were eligible for $3,500 cash voucher toward the purchase of a new vehicle. If the vehicle got 10 mpg or more the owner received a $4,500 cash back gift.

There were no purchaser income limits, but the vehicle had to cost less than $45,000. The program met with great success in 19 countries, especially Europe, where cars have better fuel economy because of high government gas taxes.

The Cash for Clunkers program was widely seen as a bailout for closing auto dealerships owned by Chrysler and GM and the auto industry. The car did not have to be made in America. The program showed that when the government is serious about getting vehicles off the road it is willing to offer direct cash payments to owners.

The Cash for Clunkers program was unfair to Clean Alternative fuel vehicles, which were not covered in the bill. For years the government has periodically offered tax incentives to clean vehicle purchasers, as much as $2,500. But the credit comes off your income at tax time, amounting to a little more than $200.

This led many environmentalists to believe Obama was not serious about cleaning up the environment. The Obama Administration settled many industry cases against the government out of Court, rather than letting the Courts dictate long and complicated policy.

Most prominently is the allowance of road construction on national property in currently roadless areas to offshore drilling in Alaska's National Wildlife Refuge (ANWR) to allowance of mountain top coal removal in Virginia, Kentucky and Ohio.

Environmentalists were clearly set back under Obama in his first two years in office. Obama offered no leadership on Climate Change Control debate in the Senate.

Another major concession in the Carbon Cap and Trade Energy bill was the continued and easier expansion of coal fired power plants. In order to gain support of moderate Democrats from coal producing states the continued focus on coal increases its consumption between 2005 and 2020.

Coal companies are allowed to buy offsets from farmers who plant trees to remove Carbon to other dubious, impossible to prove offsets. The Obama Administration took two steps forward and one step back on environmental policy. Chalk up more industry wins in the long history of battles between the environment and Big Business.

Carbon Control and Transportation Fuel

How do we bridge Conventional and Alternative Energy and Fuel? If you are a Renewable Energy purist involved in the Regional Greenhouse Gas Initiative (RGGI), the Western Climate Initiative (WCI) or the Midwestern Climate Change Accord (The Accord) then you probably feel Conventional Fuel has no place in funding from the sale of your Allowance Auction Carbon equivalent credits. But Energy Conservation suppresses oil and natural gas use by commercial and residential customers. It is fair enough to fund solar and wind Energy projects with the proceeds of the Carbon Allowance Credits.

What about transportation? RGGI does not cover it yet. It is solely a Federal purview through the United States Department of Transportation National Highway Transportation Safety Administration (NHTSA) and the Federal Motor Carrier Safety Administration (FMCSA) who set the Corporate Average Fuel Economy or CAFÉ standards for all vehicles that EPA enforces and lists on your new car sale sticker. EPA does not set the standard. They do model it in the recently released Motor Vehicle Emission Simulator model (MOVES). Transportation is mostly a Federal jurisdictional issue.

The regional agreements of the RGGI, WCI and the Accord do have jurisdiction over State, local, utility and private commercial sector fleets over five vehicles. The member states have decades old programs to promote Clean Alternative transportation fuel for fleet vehicles.

Why so little progress in this?

Biomass fuel is still experimental. It may never be produced in large enough quantities to offset more than a few percent of imported petroleum.

Electric vehicles are still a dream in some MIT and Cal Tech freshman's mind waiting for the Big Break Through in battery storage technology (a problem for solar power too). This also applies to the famous hydrogen fuel cell of the Proton Exchange Membrane.

Compressed natural gas (CNG) and liquefied petroleum gas (LPG) are two low Carbon equivalent fuels that are fossil based. CNG and LPG are cleaner than diesel for Criteria and Greenhouse Gas pollutants. CNG and LPG cars, trucks and buses are available in every Congressional District in the country. CNG and LPG vehicles are commercially available at equivalent gasoline and diesel sale price. CNG is suitable in urban counties and LPG is suitable in rural counties.

Member states of the RGGI, WCI and Accord must use 20 percent of their Carbon Allowance Auction proceeds for Clean Alternative Fuel Fleets of CNG and LPG and the construction of fueling service stations. Hybrid electric diesel is not as clean or alternative as CNG or LPG because the diesel hybrid electric vehicle operates mostly in diesel mode. Few portable emission monitoring system studies have been conducted for any vehicle, truck or bus on any fuel type.

Diesel hybrid electric vehicles need testing compared to diesel, CNG and LPG with the in use operating conditions of air conditioning, heating, hills, frequent acceleration and deceleration and heavy passenger load the same. Many of these typical urban and rural operating modes render the electric battery operation of the Diesel Hybrid Electric car, truck or bus useless. Hybrid electric battery power is hardly ever used except on extremely flat and long unsignalized roadways hardly found anywhere in urban, rural, National Park or airport transit operations.

Transportation will always be a fossil dependent fuel until battery storage and charging infrastructure is upgraded in the next 50 years. Funds are better spent in the near term on CNG and LPG cars, trucks and buses to reduce Carbon equivalent gases (Methane last several months in the atmosphere from CNG and LPG whereas diesel and gasoline Carbon Dioxide lasts hundreds of years in the atmosphere).

This may not square with the purist Renewable Energy mantras of the RGGI, WCI, and the Accord. Until battery or hydrogen power is market ready in the United States, the agreement member states are wasting investment in unproven technologies when CNG and LPG cars, trucks and buses are available in all vehicle types and models.

CNG and LPG vehicles need more fueling stations to make them convenient to the personal vehicle owner. It reduces Carbon in the near term. It reduces imported petroleum in the near term.

The agreement member states should not wait for hydrogen and advanced battery technology.

Clean, Alternative fuel CNG and LPG vehicles and fueling station construction can be built and used today in every zip code in the United States.

Chapter 2: Environmental Action Away from the Headlines

Over Fishing Regulation

Does over fishing regulation create sustainable fish yields and stable fishermen fleets? Probably it does not. But there is little other choice. The nation struggles for years to regulate Wall Street, but hastily restricts fish yields. The Democrats impose regressive taxes affecting the poor more than others, on sales, meals, hotels, alcohol and tobacco. They introduce casino gambling and new Lottery games.

Agriculture, particularly dairy farmers, another New England vital, historic industry, is in decline. Farmers can sell to land trusts (rare) or developers (more common), while fishermen must share yields, lose money, and go out of business. The government tries to offer money, but fishermen fish and farmer's farm. These are two of the deadliest jobs on earth.

The economics dictate reduce workers in these two fields. Modernization in fishing fleets with sonar and science captures more fish per trip. Then add another generation in the fishing family, and there are simply not enough fish all the time.

There will always be milk men and fish markets, just fewer of them at higher prices, two tales. It is not easy. The government usually has a young Trust Fund baby write the rules, eager to save the world, not very aware of Seaport economies. But you can not blame dreamers. You can not blame the fishermen or the fish off Georges Bank.

And what to do about the plight of the Gulf of Mexico fishermen devastated by the oil spill in 2010? BP better still be paying them now. This was very untimely and unfortunate for the Gulf of Mexico economy and also to the oil industry.

Environmentalists squirmed in their seats accepting limited expanded coastal oil drilling. Now everybody is against it. And nuclear, another bitter pill for environmentalists to accept on the Obama agenda to get the Energy bill passed. There will be a Chernobyl on U.S. soil. Just wait. Most of the existing spent nuclear fuel is poisoning local wells, ponds and rivers all over the land and no new funds for clean up.

And coal. Where did they go? This is another deadly job for low-skilled, uneducated and high risk workers in rural counties for dirt cheap pennies, tons nonsense Carbon Energy. The coal folks are hiding after the disasters all over the world in their old world, outdated, shoddy industry with very little benefit to the worker or the environment. The local towns in the rural Midwest from Virginia to Missouri need their jobs, so they pay with their health.

What do you pay with, Mr. Middle Class? What crappy new, high risk jobs can we give the undereducated lower class? What new tax can we further widen the gulf between classes. Oh yeah, the Carbon Tax. Geesh, one volcano can ruin my European vacation and create more Carbon, Sulfur and Nitrogen pollution than man has ever made. Go figure.

Land Use

America has witnessed great suburbanization in the last 60 years. This prompted what some call sprawl or automobile friendly development, great for economic and personal mobility. But, as really urban states (we are a mostly an urban country despite vast tracts of undevelopable and potentially developable real estate) have vast tracts of below built cost value housing, new strategies from the cities or re-development need evaluation.

Most think all land from Maryland to Maine is built out. But many central city areas have been redeveloped numerous times in their some almost 400 year history since settlement. In the 1960's, as awareness of environmental impacts from land use and transportation became more commonly thought of, Planners, mostly architects, real estate attorneys, developers and state, county and town Zoning and Development Officials thought of new ways to save and create open space and reasonable land use patterns.

Land use planning and regulation is a neighborhood to town, tract by tract, local issue. There are established entities in this market space. Many have vested interests. But as towns found housing, social services for families, like schools and libraries, too expensive to build and maintain, they looked to commercial development to lower the average tax rate and raise revenue for public services.

In more dense areas, Planned Use Developments and density bonuses (providing low market rate rental or sale price units gets the developer extra than allowed zoned units) for more compact development by developers, were passed by town and city officials. The South and Southwest might want to consider this re-development option for built or undeveloped suburban tracts. Usually a local shopping plaza with a grocery store, pharmacy and other basic needs like banking, beverage stores, boutiques and entertainment are also included

Now we need to link a sustainable local transit shuttle paid for by communal developers to provide some type of fee for service or discounted paratransit for the transit dependent population, transportation option. This is reasonable as many citizens become aware of pollution and infrastructure constraints caused by the personal auto.

Coal the Old Pollutant that Remains a Problem

As scrubbers become common, coal air pollution is flushed into rivers. Coal plants in the U.S. are finally installing particulate matter scrubbers to reduce air pollution. But now the solid distillate of the air waste is flushed into rivers.

Plant managers at Hatfield's Ferry in southwestern Pennsylvania say they have installed equipment to reduce toxic pollution from the coal scrubbers discharged into the Monongahela River. EPA estimates 50 percent of all U.S. coal plants now have scrubbers on them.

There are no Federal rules on coal plant air scrubber toxic release into rivers, landfills and ground water. Toxic chemicals like arsenic and lead are unregulated. Enforcement of outdated Clean Water Act standards is alarmingly lax. 90 percent of 313 coal power plants broke the Clean Water Act law since 2004. Those fined paid very little in penalties. Hatfield's Ferry broke the Clean Water Act 33 times since 2006. The plant owner only paid $26,000 in fines while at the same time the holding company earned $1.1 billion.

"We know that coal waste is so dangerous that we don't want it in the air, and that's why we've told power plants they have to install scrubbers," says Senator Barbara Boxer (D-CA) Chairwoman of the Senate Committee on Environment and Public Works. "So why are they dumping the same waste into peoples' water?"

EPA said early in the Bush Administration it would issue new power plant waste rules. It said the same thing again when a Tennessee coal was dam ruptured and flowed 1.1 billion gallons of coal waste through houses and farms in 2008. But state regulators often fight Federal standards at the sway of utility lobbyists.

Hatfield's Ferry is in a lawsuit filed against it by local residents. Plant managers say there is no problem since they put in a $25 million water treatment facility to take out the toxic particles and solids from the coal scrubber waste. The solids are buried in a plastic lined landfill. The managers say limits on arsenic, aluminum, barium, boron, cadmium, chromium, manganese and nickel are not needed since they are unlikely to harm the Monongahela River and exceed contamination levels.

"Allegheny has installed state of the art scrubbers, state of the art wastewater treatment and state of the art synthetic (landfill) liners," the company wrote. "We operate to be in compliance with all environmental laws and will continue to do so."

"It's really important to set a precedent that tells power plants that they need to genuinely clean up pollution, rather than just shift it from air to water," says Abigail Dillen lawyer with EarthJustice representing local environmental civic groups in petitioning the Pennsylvania Court to strengthen rules on Hatfield's Ferry. Dillen is arguing for "zero discharge" rules, more expensive but capable of eliminating virtually all pollution. Pennsylvania regulators say they have good water pollution rules for Hatfield's Ferry and landfills.

"We asked the plant for estimates on how much of various pollutants they are likely to emit, and based on those estimates, we set limits that are protective of the Monongahela," says Ron Schwartz, a Pennsylvania regulator. "We have asked them to monitor some chemicals including arsenic, and if the levels are too high, we may intervene."

These problems are not unique to Hatfield's Ferry. 21 power plants in states like Alabama, Kentucky, North Carolina and Ohio have polluted rivers with arsenic levels 18 times higher than the Federal drinking water standard according the New York Times review of EPA data. In Florida, Illinois, Indiana, Maryland, North Carolina, Ohio, Wisconsin and elsewhere power plants have polluted rivers with other chemicals at unhealthy levels. None of these plants have ever had penalties for that pollution. Discharge permits were not adjusted to prevent future pollution.

Lobbyists long held sway stopping EPA from issuing new environmental rules under the Bush Administration. This changed under Obama as he enforces existing environmental laws and prepares toughened, new rules.

Mountaintop Coal Removal

In a seldom used power, EPA blocked a mountaintop coal removal project in Logan County, West Virginia. The mine already had a legal permit to operate given in late 2007. This was a dramatic move by the Obama Administration. EPA has checked dozens of previously issued permits in West Virginia for mountaintop coal removal projects. EPA approved one large mountaintop coal removal project in January 2009.

Under the Clean Water Act of 1972, EPA has the power to over turn permits and block actions that have an adverse impact on a waterway. The special EPA power has only been used a dozen times in 38 years. Spruce Number 1 will bury 7 miles of sensitive streams in Logan County, West Virginia. The toxic debris from the removal of the mountaintop will dump toxic material into streams that will harm aquatic life. The region is one of the world's richest collections of salamanders.

The mine was given the permit to operate in late 2007. Law suits against the project caused EPA to reconsider the permit. In 2009 EPA started talks with Arch Coal, the owner of the mine from St. Louis, to modify the permit. EPA finally said the talks failed. Arch Coal says it would proceed with all legal means to continue the project.

Scientists urged the ban for mountaintop removal for coal mining. A group of scientists from the University of Maryland urged the Federal government to ban all permits for mountaintop removal for coal mining.

But the scientists took the unusual next step and entered the political fray by publicly advocating the ban.

Scientists reviewing public policy are supposed to be objective, neutral and dispassionate, making a conclusion and recommendation in writing, but not becoming public policy advocates for one side or the other.

When this happens its damages the credibility of scientific review process, no matter how deleterious the impact from the practice studied are. In recent years after the environment suffered in the last decade, scientists see the Obama Presidency as a time to catch up on environmental regulation and enforcement. Industry has been quick to react but somewhat powerless after eight years of voluntary reporting and compliance.

EPA issued a statement reinforcing the scientist's findings noting that the report, "underscores EPA's own scientific analysis regarding the substantial environmental, water and human health impacts of these mines".

Industry claims that after the mountaintop is finished, the damage to nearby streams is only short term, not lasting more than 18 months.

In the report, the scientists said the damage could last hundreds of years. The report said over 1,500 miles of streams had already been destroyed in West Virginia, Kentucky, and Tennessee.
Mountaintop removal involves the peaks being sheared off with heavy machinery and explosives, allowing quick, easy and cheaper access to the coal seams inside. The excess rock and dust gets washed by rain carrying sulfur and toxic heavy metals like cadmium and mercury into nearby streams, harming insects and fish.

The lead scientist said the evidence against mountaintop removal is as clear as the link between smoking and cancer. People in the region are also at risk from the toxic air dust and well water contaminated with chemicals from the mines and ingested fish laced with toxic metals.

EPA then essentially immediately banned and halted a wide spread practice of mountaintop coal removal in remote Appalachian coal regions of Eastern Kentucky and Western West Virginia. The rules were a rare EPA move and went Final immediately once issued.

The practice known as mountain top coal removal occurs in isolated, rural Appalachia where the coal industry is the only major employer.

The coal in these parts of Kentucky and West Virginia is thin and close to the surface. Conventional coal mining drilling does not work in this area the size of Rhode Island. The mountaintop is blasted off and the coal easily and economically harvested, employing hundreds of local workers in several counties in both Kentucky and West Virginia.

Metropolitan Northern Virginia, Washington, D.C. and Maryland region powered by PEPCO relies to a great extent on power from these mountains. Nationally, the coal is 10 percent of the annual coal yield.

President Clinton moved to halt the practice, but President Bush reversed the rule, granting over 500 permits to the mountain top coal removal industry filling in streams 176 miles long with arsenic, cadmium, cobalt, benzene and polyaromatic hydrocarbons, all proven carcinogens, into community wells and drinking water. Well water once clear in Twilight, West Virginia is now tainted grey with a pungent odor and acrid taste.

In addition to increasing early and preventable human morbidity and mortality rates, a rare salamander on the Endangered Species List inhabits the stream in one of the few locations it is found.

The coal industry vowed to fight.

Mining Industry

Miners must pay cleanup costs. In a Court ruling in February 2009 the Sierra Club won a case forcing EPA to adopt rules to regulate mining waste. By filing bankruptcy, mining companies were allowed to wiggle out of waste cleanup costs. EPA is beginning the rule making process with hard rock mining companies (uranium, molybdenum), but will follow with rules for other types of mining operations. The proposed hard rock rule will be published by EPA in the spring of 2011.

The National Mining Association says that miners are already covered by other Federal and state laws and that "EPA ignored critical facts and data". The Senate held hearings on making miners pay royalties to the Federal government dating back more than 100 years and totaling over $245 billion. Mining companies can buy public land for as little as $2.50 an acre, even during the Great Recession. An example of one such company is Asarco, which filed for bankruptcy in 2005. Asarco owned 94 Superfund sites in 21 states.

Drilling on Public Lands

Interior Department issued new guidelines for drilling on public lands. Secretary of the Interior Ken Salazar released new government guidelines for drilling on public lands.

The guidelines are meant for existing and sought after permits from oil and natural gas companies to drill in the National Parks and protected areas and public lands owned by the Federal Government.

In 2009 Salazar suspended 60 of 77 leases in Utah that were approved in the last days of President Bush in office. One percent of oil and natural gas leases on public lands were contested in 1998. Now the figure is around 40 percent.

The new guidelines are meant to streamline the leasing process so there is more transparency, clarity and less litigation in leasing public lands to the oil and natural gas industry.
The government pays millions of dollars in legal fees defending its decisions on the approval or denial of drilling rights. Salazar is setting up a process that will be out in the open and give industry a clear set of rules to comply with, hopefully, lessening litigation and lengthy delays for industry, while protecting public lands and National Parks.

Salazar's decision is an indication how he will approach the politically charged issue of drilling in the Alaskan National Wildlife Refuge (ANWR), off the Santa Barbara and California Coast, and off of the Florida Coast and up the Eastern Seaboard. George's Bank has already been protected and placed off limits due to its fishing economic vitality to New England.

It is clear the country has vast reserves of oil and natural gas in these protected and sensitive ecosystems. But the U.S. needs more domestically produced offshore oil and natural gas, where we have vast untapped reserves.

Interior reviews several different aspects during an environmental review of a drilling project, the economic, environmental, ecosystem, cultural, historical, safety, security, and mobility impacts on industry, the public interest, environmentalists, fishermen and Native American tribes, all with valuable, and important self interest on one side or the other, depending on the particular case.

On the North West Coast salmon have been suffering from dams built in the last 100 years. Many are coming down in northern California and southern Oregon to revive salmon stock on the Klamath River. This is due to a Salazar ruling late last year, reviving river salmon fishing industries and the Native American cultural importance of the salmon.

Look for Salazar and Interior to take a moderate approach to balancing out the needs of industry and environmentalists as a workable solution to offshore drilling comes up in the coming years.

Logging in Oregon

Logging limits back in place in Oregon. In early 2009 Interior Secretary Ken Salazar reinstated limits on logging in old growth forests in western Oregon. The Bush Administration had doubled the amount of logging permissible in the habitat of the spotted owl. The compromise first reached 15 years ago protects not only the spotted owl, but watersheds, trout and salmon fisheries, and other endangered birds.

President Bush would have allowed timber companies to log up to 502 million board-feet of lumber annually from 2.6 million areas of forests in the region. This is double the amount contained in the 1994 Clinton brokered Northwest Forest Plan. Unemployment in Oregon is 12.1 percent in the midst of the Great Recession, one of the highest in the country. Douglas County, where forests involved are located, unemployment has reached 16.9 percent, mostly due to the closing of sawmills and loss of timber and logging jobs. Tom Partin, president of the American Forest Resource Council said the Bush plan, "would have given our timber-dependent communities a real boost".

Environmentalists argued that the Interior Department's Bureau of Land Management, forest overseer, did not consult with the Fish and Wildlife Service about logging's impact on endangered and threatened species. The Bush plan would have reduced protected habitat of the spotted owl, an endangered species.

PacifiCorp Removes Four Dams on Klamath River

PacifiCorp agrees to remove four dams on the Klamath River in Oregon to protect salmon. The Klamath River dams in Oregon were constructed between 1918 and 1961 on the upstream portion of the Klamath River. Native Americans, fishermen and environmentalists cite large declines in the salmon population and economic dislocation. Managing the tremendous resources of the Klamath River has been one of the biggest water resources challenges in recent times. The Klamath River was once home to one of the world's largest salmon population.

The fight over the Klamath water is mainly due to the dams preventing upstream spawning by the salmon. Farmers have lobbied for more water for irrigation and utilities for electric power generation. The cost of the dams removal is placed at around $450 million. $200 million will be paid by a small fee on PacifiCorp's customers, a largely Oregon base, the remainder will come from bond revenue from California, which also has a large interest in the Klamath River salmon.

The Interior Department is in the middle of an Environmental Impact Assessment to determine the full cost of the dam removal and the impact on fish populations, as required by Federal law. The Interior Department has until March of 2012 to decide about whether the dams should be removed. Balancing competing interests of Native American's, fishermen, farmers and utilities economic, social and environmental costs and benefits will be addressed by the Interior Department.

PacifiCorp agreed to remove the J.C. Boyle, Copco Nos. 1 and 2 and Iron Gate dams because the company felt is was an environmentally wise business decision based mostly on economics. PacifiCorp is a subsidiary of billionaire investor Warren Buffett's Berkshire Hathaway Empire. The group was looking at lengthy and costly litigation and re-licensing requirements for the four dams. The oldest damn was built in 1918.

"As a utility we don't typically take dams out," said PacifiCorp's Dean Brockbank. "We have achieved an agreement that is in the best interest of our customers - the lowest cost and risk compared to the alternative."

Chapter 3: Offshore Drilling and the Deepwater Drilling Gulf of Mexico Oil Spill Catastrophe

Drilling off of Alaskan National Wildlife Refuge (ANWR) in Alaska ends. It never actually got started. The Bush Administration did an end run around the Congress by signing an Executive Order allowing the Interior Department to lease land off of Alaska near the Arctic National Wildlife Refuge (ANWR) for oil drilling. A Federal Appeals Court in Washington, DC suspended the five year program that began under Bush in 2005, declaring it needed further environmental review of the impacts of oil and natural gas drilling on ecosystems in the outer Continental Shelf, not just the Alaska coastline.

Inupiat Eskimo fishing communities and environmental groups hailed the decision. The oil industry said America can not wait to tap the estimated 90 billion gallons of petroleum off Alaska's north coast in a time of imported Energy dependence, fluctuating and now again rising, at the pump gasoline prices.

Gasoline prices are up sharply as summer approaches but still well below last summer $4 plus per gallon gasoline prices. The petroleum off Alaska is estimated by industry to contain more oil than Mexico, Nigeria and Kazakhstan combined.

The Interior Department has also halted leasing off of the potentially oil rich California Coast, Atlantic Coast and parts of the Gulf of Mexico.

Ken Salazar, Secretary of the Interior is having the department hold extensive public hearing in the next several months in several states to get comment on offshore drilling in America's waters.

Public opinion may have changed since many of the bans were imposed. If gas prices spike again this summer, offshore coastal drilling will be more open to the public. In addition, oil and natural gas drilling technology have made great improvements since offshore drilling was banned a few decades ago.

Interior's decision to halt offshore drilling while further environmental review takes place for the ANWR area and solicit public comment on the California, Atlantic and Gulf Coast regions is a good move.

Even Obama conceded to a degree during the campaign that more domestic oil and natural gas exploration is needed. It is unclear what his stance will be now on offshore drilling. Certainly, if oil prices spike close to $4 again this summer, pressure will be on the Obama Administration and Congress to allow more if not only some offshore coastal drilling.

In 2008 we were paying close to $30 more per tank to fill our cars, in the summer of 2009 about $5 more a tank. If prices spike, Clean Energy will get more private investment.

Private investment in Clean Energy collapsed after oil prices fell at the pump last year and the Great Recession began. Let industry make its case on the impacts of new exploration and drilling technologies and its impact on the environment.

A key Federal agency backs offshore drilling restrictions. The Interior Department is revising Bush era plans to drill for oil offshore in Alaska, California, the Gulf of Mexico and the Atlantic Coast.

The National Oceanic and Atmospheric Administration (NOAA) issued strong testimony urging a ban on offshore oil drilling. The testimony calls for a buffer zone around the Southern California Ecological Preserve off Santa Barbara, an all out ban on Arctic drilling in Alaska and a one year delay on the Interior's 2010-2015 drilling plan until the Obama Administration's Ocean Policy Task Force finishes its work.

In Alaska's North Aleutian Basin and the Chuckchi Sea, NOAA is "very concerned about potential impacts to living marine resources and their habitats, viable commercial and recreational fisheries, and subsistence use of marine resources as a result of future lease sales, exploration and development".

Dr. Richard Steiner of the University of Alaska says, "The significance is that here we have one Federal agency supporting what we have been saying all along regarding the push to lease offshore in Alaska." He says NOAA's testimony has put "Interior in a corner in all this."

Jeff Ruch, Executive Director of Public Employees for Environmental Responsibility, sternly warns, "If NOAA's warnings are not heeded, Interior's offshore leasing plans will again be ensnarled in litigation."

Representative Doc Hastings (R-WA), the lead Republican on the House Natural Resources Committee blames the Obama Administration for levying "a de facto ban on offshore drilling."

Interior Secretary Ken Salazar says the future leasing plan, "must take into account several key considerations, including area's of the ocean that are critical for military training and the nation's defense; other economic benefits of the oceans, including fisheries, tourism and subsistence uses; environmental considerations; existing oil and gas infrastructure; interest from industry; and the availability of scientific and seismic data."

The Interior decision will impact every mile of U.S. shores and result in litigation, regardless of the final details of the 2010-2015 U.S. Offshore Drilling Plan.

Obama was set to allow offshore drilling in Alaska. In the spring of 2010 the Interior Department gave Shell Oil the right to drill two test wells in the Beaufort Sea of Alaska. The Obama move indicates a willingness to work with industry on offshore oil drilling in environmentally sensitive areas.

Bush had gone forth with a major Executive Order late in his term but the Courts and Obama put offshore drilling in America's coastal waters on hold earlier in 2010 until a full environmental, biological, ocean ecosystem, cultural and economic impact assessment could conclude to assess the impacts of unlimited offshore oil drilling.

Drilling off the American Coast was common through much of the 20th Century until the environmental movement of the 1970's and oil spills. Environmentalists successfully blocked offshore drilling for a generation. But as the country faces increased needs to get off imported petroleum from the Middle East and Venezuela offshore drilling is getting Obama's attention. Obama is wise to offer a gift to the oil industry. The current Climate Change legislation on the Hill greatly costs and hurts the oil industry.

Oil companies say their drilling technology has vastly improved in the past generation. It is less invasive, safer, and cleaner and fewer derricks are needed, preserving and enhancing the coastal visual scene. Environmentalists seem unwilling to accept some modification of plans for offshore drilling to allow some limited drilling in areas found to pose the least risk. Environmental groups and industry have learned a lot from their Court squabbles in the last 50 years. Both sides are increasingly working together on joint programs to clean the environment while maintaining sustainable economic growth.

Even Libertarians and Republicans were against offshore drilling in the election summer of 2010 when the BP Deepwater Oil Spill happened. We all feel good to pick on BP, in the 24/7 media hype. Ugly details no one really cares about or will a year from now when the wild life and fishermen will still suffer. I know you feel good to harass an apparent bad corporate player. This is the typical Post-Modern reaction of the masses in the West and world over.

So are we going with Green Energy? Do we abandon fossil fuel and nuclear Energy? No. Not even massive suburban transit can stop you from turning the ignition and flipping a switch for light. No. Might as well admit the U.S. is addicted to oil, coal and nuclear power. The Energy sources will remain the key economic ameliorative to the World Economy for hundreds of years to come.

Only Big Oil, Big Coal and Big Industry, the status quo per the usual, huge employers of both suppliers and local salesmen, have the economic muscle and risk capital to nourish the 50 year old Clean Technology industry.

Some individual investors can play a role in the new Energy Technology, not the average investor. As oil costs at the pump look to go beyond $5 a gallon in 2011, look for fuel diversification to stay in the lens of cable media. Mass media of Big Press can not make much money on mundane, but vital Energy policy, so look for the usual dribble of Big Networks, Big Cable and Big Press sensationalism coverage. It is their standard business model.

EPA approved chemical dispersants in Gulf Oil Spill clean up in risky move. These detergents were untested. The wildlife, ecosystem and habitat impacts were unknown. This was similar to early Chemo therapy by using a poison as cure. Desperate were the times in the Gulf.

What are these chemical dispersants? High school chemistry says they are just tallow or soap based pumped up detergents, like you find in your gasoline in your car to reduce negative impacts of dirty petroleum.

The negative impacts should not harm the environment or human health, but this is unknown.

It is a shame we have to resort to this desperate measure. There was little other choice, except continued known harmful impacts on the ecosystem, economy and health.

BP apparently had several plans they attempted to use to stop the fossil fuel flow over and about four five back up plans. Hopefully the first one works and they do not have to go down a laundry list of trial and error.

Does the this spill negate the need for offshore oil drilling? No. Just like Three Mile Island and Chyrnobyl did not ruin the nuclear industry. But the hyper responsive media types and activists have swayed politicians in Coastal States and Congress to stop all new drilling. This is temporary, but it could last for decades.

But the only economically viable area to explore and drill for new oil and natural gas fields is off of Northern Alaska in waters opened by sea temperature rise and Global Warming.

All the hype, chest pounding and puffery over continental offshore oil drilling is a waste of time for publicity and ambulance chasers in the media to profit off the sad tragedy of the Gulf Oil Spill.

Disasters, waiting to happen. Manmade and natural. They are always a tragedy. The immediate human lives lost. The long term social, environmental and economic damage. Disasters are hard to predict, hard to prevent, hard to prepare for and even harder to respond to in an efficient and successful manner.

In 2010 we had the Earthquake in Haiti and several others around the active Pacific Rim. Mining disasters in the U.S. China and Russia. The oil spill in the Gulf. The volcano in Iceland.

Some of these are implicated by Climate Change, others add to the problem of Climate Change.

The BP owned and managed Gulf Oil Disaster, greater than the Exxon Valdez, has killed all new permits for off shore oil drilling. The long term prospect looks dead also. Perhaps as the memory recedes, economic drilling in the Eastern Gulf and Northern Alaska will happen, but not anytime soon.

And we have not had a nuclear disaster in the U.S. in nearly a generation since President Carter saved Three Mile Island in Harrisburg, PA.

Not really anything to comment on or report. I am just mourning the loss of human lives common in the Energy industry, our need for diverse Energy sources, tried and true and new and blue, and of the course the ecological, and habitat, and wildlife, environmental and economic damage.

All very sad.

We are merely human and mistakes happen. Mother Earth will do her thing on her on cycle. Today we shout and scream. Executives should take responsibility. They are acting like teenagers caught with a six pack. Air traffic control in Europe is under fire for too much caution. Always blame, no matter the response. We need responsibility and accountability. But it seems like controlling the news spin cycle is all responsible groups care about.

The Obama Administration officials speaking of the common frustration over the Deepwater Oil Drilling Platform Disaster in the Gulf of Mexico, off of the already ravaged Louisiana Coast, this afternoon used unusually harsh and stern language against the BP Team responsible for the incident. No answer yet.

The rupture polluted from April through August 2010, five months with a magnitude far worse than the Exxon Valdez super tanker oil disaster in fertile Alaskan waters 30 years ago. The Obama Team sent the right signals. No words of punishment or criminal charges, yet. The BP Team, while offering numerous, unknown and never tried stop gap measures, stumbled badly before the Congressional inquiry.

This is the kind of old industry response ruining many Corporations in obfuscation and fury as public reaction with the purse batters future sales and growth for decades.

New Media is spreading the news about what a poor neighbor BP is all over the Domains. Citizen response solutions should roll out next, technology failing thus far with deadly detergents referred simply as dispersants, as if safe for dishes and laundry.

Why not try non-toxic solutions like pet and human hair in nylons? The citizens have found just as many if not more, solutions to stem the oil hemorrhage as science and engineering. Oh, no one will make money from it, if it even works an ounce as much as any of the so far useless modern tech band aids.

Boycott, last heard against meat 40 years ago due to the now weekly occurrence of manufacture contamination, was in fashion in 2010. The old citizen brigades were out in full force on the Domains without the need of Big Non-Profits or any one leader. Social democracy here, right where it started and transplanted to China and now Iran.

Social democracy can not face suppression or torture or editing or a slick fund raising campaign at the wrong hour. I say honk your horn every 4 hours or at least at an intersection, when you are not annoyed, just disgusted at Big Business contempt for compliance with the law.

Government response is critical. The fine at every level of government for loss of human life, pollution, damage and threats to safety, mobility and security of the world wide vital Gulf Fishing Fleet is $25,000 per day until to contagion ends. Enforce it immediately.

Clean Energy and Oil Accountability Act of 2010

Senator Reid (D-NV) introduced the Clean Energy Jobs and Oil Accountability Act of 2010 to the Senate in August 2010. It had strong bi-partisan support first aiming at preventing another Deepwater Horizon Oil Spill catastrophe and protecting natural resources, workers who rely on natural resources for living and coastal communities' shorelines.

-    Removes $75 million cap on corporate liability
-    Waiting period for payment of claims is reduced to 30 days from 90 days
-    The Clean Water Act power is given to the President to delegate to the Executive Agencies the powers to prevent injury to the economy, jobs and the environment with major focus protection of private property income loss, injury, damage or permanent loss of use. It expressly allows the President to authorize new regulations to protect the loss of profits or earning capacity.

The Act has 5 major Divisions.
1. Oil Spill Response and Accountability.
2. Reducing Oil Consumption and Improving Energy Security.
3. Clean Energy Jobs and Consumer Savings.

4. Protecting the Environment and

5. Fiscal Responsibility.

Title 1. Removal of Limits on Liability for Offshore Facilities.

Title 2. Federal Research and Technologies for Oil Spill Prevention and Response

Title 3. Outer Continental Shelf Reform

Title 4. Environmental Crimes Enforcement

Title 5. Fairness in Admiralty and Maritime Law

Title 6. Securing Health for Ocean Resources and Environment (Shore)

Title 7. Catastrophic Incident Planning

Title 8. Subpoena Power for National Commission on BP Deepwater Horizon Oil Spill and Offshore Drilling

Title 9. Coral Reef Conservation Acts Amendments

Title 20. Natural Gas Vehicle and Infrastructure Development

Title 21. Promoting Electric Vehicles

Title 30. Home Star Retrofit Rebate Program

Title 40. Land and Water Conservation Authorization and Funding

Title 50. National Wildlife Refuge System Resource Protection

I propose an Amendment to the Senator Reid "Clean Energy Jobs and Oil Accountability Act of 2010". I propose adding in Division B Reducing Oil Consumption and Improving Energy Security a Title XXI. Please Senators insert "Propane Vehicles and Infrastructure".

I have close ties to Maine, New Hampshire and Vermont. Many of my friends and family use propane for heating, cooking and hot water needs everyday all year long. Why not as a fuel for their Ford F-150 or 350? School buses run on propane. Regional Budweiser trucks could easily run on Liquefied Propane Gas in rural areas of New England and the whole country.

I call on Senators Snowe (R-ME), Senator Gregg (R-NH) and Senator Leahy (D-VT) to introduce Title XXII Propane Vehicles and Infrastructure to the Act under Division B Reducing Oil Consumption and Improving Energy Security of the Clean Energy and Oil Accountability Act of 2010.

Propane is by-product of petroleum and natural gas refining. It is not a natural resource. Propane is neither a liquid like petroleum nor a gas like natural gas. Propane is both a liquid and gas, depending on the amount of compressed pressure it is under and temperature.

Propane vehicles have the same driving range as natural gas vehicles. Both propane and natural gas have the same driving range and like any fuel its natural capacity is matched by tank size. Modern tank configuration for all fuels conforms to body specific contours of extra design space in the under carriage for each model. Diesel long haul trucks we all know have numerous fuel tanks to carry our goods across country.

All fuels are either liquid or gas or move back in forth in a transient state. Liquid fuels are unsafe in an incident because they pool on the ground and can ignite from sparks. Gaseous fuels vent to the air in a spill or accident and can also ignite with sparks. All can spontaneously combust under varying conditions. All Clean Fuels and Traditional Fuels have matched driving ranges, performance and safety. Fuel usage or efficiency, as with anything, varies, thus the need for bigger tanks for some fuels.

In the past 20 years we have seen market barriers to Natural Gas and Propane Vehicles in terms of pricing of new vehicles, fuel stations and town, county and state laws barring brought up to par with gasoline and diesel.

In the past the Senate and House have provided tax incentives or tax credits of $2,000 to $5,000 depending on vehicle size (car or truck). These are tax credits and amount to a small $200 per vehicle, not enough to overcome the market barrier. The Congress must allow a straight, flat, progressive after tax computation tax deduction of $2,000 per incremental purchase price for a propane or natural gas car or truck.

In addition, the $250,000 fuel station construction tax credit must be made a tax deduction. If the Senate and House are serious about boosting the Alternative Fuel car and truck industry, promoting Energy diversity and use of local labor, skill and resources, I suggest you consider my proposition.

Chapter 4: Clean Water Act and Major Environmental Protection Acts

The Clean Water Act requires Congressional action. President Nixon established the Environmental Protection Agency around 1970. Ensuing Democratic Congress' regulated Clean Air, Clean Water, HazMat, Toxic Substances and Solid Waste. Only Nixon could accomplish this, a deft politician who learned from the social protests of 1960's in this area.

The Democrats lost control of both House and the Senate in 1994. The Democrats were kicked out of two generations of Liberal progressive, social agenda, Big Government and centralized power. Nixon augured in the Conservative Era that will last two generations, strengthened by President Reagan's bold de-regulation that effectively changed logistics and inventory, reviving the economy. Reagan also changed Federal government staffing, now mostly out sourced to temporary contractors, who often try to become government employees. President George H.W. Bush and President Clinton moderated back right then left towards the center.

In 1994, Newt Gingrich won power from the Democrats with his Contract On America. Many of the Federal Congressional environmental acts were passed in the late 1970's and 1980's with sunset expiration dates. All have expired except Clean Air.

The Clean Air Act was significantly amended by Congress on November 15, 1990 and signed by President George H.W. Bush. The environmental group leaders negotiated tight six month, 12 month, 18 month and three year deadlines, with loss of Federal Highway Funding for failure to meet planning deadlines, not the Federal Standards.

Great progress has been made on the Clean Air standards. Many areas have attained and maintained the Acid Rain and Carbon Monoxide standards. Ground Level Ozone in hot summers and local Particulate Matter (soot) remain a problem but are much better. Cancer rates are down for many reasons since 1990.

Interestingly enough, President-Elect Clinton gave the environmental group leader's their wish and appointed most to key EPA leadership positions to implement the nearly impossible deadlines they created in lobbying. "Be careful what you wish for", Mary Nichols, now the Executive of the California Air Resources Board, said in the winter of 1993 to Washington, DC Transportation Officials when she was at EPA.

The Clean Air Act and why it stays in the news, is the only major Congressional environmental act with no sunset expiration. The EPA has authority to review, regulate and implement, with three year pollution report progress benchmarked and program changes made.

This is the model we need for Clean Water, HazMat, Solid Waste, and Toxics. Benchmarked, clear achievable reduction and re-use goals, with sanctions for failing to plan three year progress reports, regulatory review and revision and implementation by the states and local agencies.

In the specific case of the Clean Water Act, nearly half of all major polluters are exempt because of Supreme Court rulings over the phrase, "discharge into navigable waters". For years this was broadly interpreted to cover every body of water, but stricter interpretations exempt inland ponds and streams, swamps and water ways that dry up or are not navigable. Thousands of EPA violation enforcement cases have been dropped because of this gap. Clearly, the time for re-authorization with the no sunset expiration dates is required by Congress soon.

Several key U.S. environmental legislative acts have expired in the last two decades. The Clean Water Act. The Toxic Substance Control Act (TSCA). The Comprehensive Environmental Reclamation and Cleanup Act (CERCLA) more commonly known as Super Fund. The Resource Conservation and Recovery Act (RCRA). All expired under President Clinton and President Bush. The only functioning Congressionally legislated environmental protection act still functioning, largely because of no expiration date and funding by users, is the Clean Air Act Amendments of 1990.

The Administrator of the United States Environmental Protection Agency, Lisa Jackson, Massachusetts's Senator Kerry and Senator Brown and Congressional Representative Stephen Lynch are non-responsive to inquiries on the status of these four major environmental protection legislative laws.

Some regulations remain in effect at both the Federal and State level. Funding is down to less than 1 percent. User fees are mostly uncollectable or uncollected. Enforcement of environmental law and regulation is virtually non-existent at all levels of government for the past, going on, 20 odd years. Public elected and appointed officials no longer care about the old news of environmental protection. Job loss, economic impact, expensive compliance, operating and enforcement costs make environmental protection unregulated, unfunded and unenforced.

Some regions of the country pay more lip service to it than others. All levels of government share the burden, responsibility and negligence for failure to protect the environment, public health, private property, wildlife and habitat, sensitive cultural resources and the professional industries working in environmental protection. Zero media coverage. Zero statements from environmental interest groups and prominent government officials. Publicity zero, except when boil water orders are in effect, a coal mine or oil rig explodes.

Then the typical media spot light, outrage and sensationalism from the Left, the Right and every body trying to make a buck off of the latest manmade or natural catastrophe flourishes for a week or so.

Twitter, Digg and Facebook buzz. Until Lady Gaga or some other public media darling, grabs our attention with a baby and rumored marriage.

Now it is Climate Change, with a name and focus switch to Energy. It's new. It's sexy. It's hip. Green design. New technology. New investment. New industries. New jobs. New research. New same old hype with no action from the Left and shouts, undocumented, of harm to the precious economy from the Right.

It was not simply high oil prices in the 1970's and international tension that slumped the U.S. economy. U.S. manufacturing plants were old and outdated, non-competitive in the emerging world market. Japan and Germany, rebuilt by the Marshall Plan in 1950's, had new modern, industrial production facilities absent in the U.S.

President Nixon is credited for authorizing numerous Federal environmental protection acts, including water, air, toxics, waste management both past and present, and safety. Nixon regulatory policy, carried on by House Speaker Tip O'Neil (D-MA) after Nixon's crimes, re-engineered American process design, production and distribution in the U.S.

By the 1980's President Reagan could claim victory for the economic boom that would be enhanced by air re-authorization by George H.W. Bush in 1990, and boosted by continual technology explosions on the market that brought the average work desk a computer, the home a microwave, the cell phone and mobile technology.

But the auto industry boom of the 1980's was unheralded. The fuel economy and Criteria Pollutant (VOC, NOx and Carbon Monoxide) tailpipe controls nearly ruined Chrysler but catapulted Chevy and Ford to continued market dominance staving off the Germans and Japanese.

Now Climate Change Control Cap and Trade legislation, the mere threat of it, is re-engineering the American and world economy, hurling China ahead of the West and the leader of the old Colonized States to an Emerging World of unified development and goals.

China is now a key U.S. partner, militarily reigning in the aging North Korea, and spurring finance of the Western free market bail out while opening its door to a new floating world currency market, no longer in fear of subjugation to foreigners. China's vast manufacturing muscle sells basic goods of luxury to the Emerging Nations and promotion of fierce competition with the West in new technology. This is happening all despite the typical hi-jinx of corporate espionage and market research (i.e. polite pirating).

So where do we go with a floating world market that must cut value and raise revenue? We work, like Spartans to vacation in Athens, and stave of the encroaching empire. Belt tightening is never easy. The discipline to shut a plant down because the Carbon Cap is reached requires new solutions in all technology not even on the minds of inventors yet to stave of plant closure. The New Clean Economy is transforming East and West to a new, brighter future in a clean, safe, secure, sound world.

Toxic Substances Control Act (TSCA) Set for Re-Authorization.

After lapsing the Toxic Substances Control Act (TSCA) is set for re-authorization by Congress soon. The TSCA lists chemicals that pose a human health hazard. The main public awareness of the TSCA is the Material Safety Data Sheets or MSDS. The MSDS lists the chemical component, symptoms of exposure, contact numbers and emergency treatment guidelines.

But there is a loophole in the law that allows corporations to claim trade secret privacy of a chemical. Only a handful of top government scientists have access to the hazardous chemical formulations. State and local Hazardous Materials Response teams and public health officials do not have access to the trade secret chemicals.

Nearly 20 percent of all chemicals are protected by the trade secret status in the TSCA. Approximately 700 new chemicals are brought to market each year. The program under the original 1976 Toxic Substances Control Act worked well at first. But now after 34 years, the secrecy exemption is being abused. Nearly 95 percent of new chemicals each year request the secrecy exemption, according to the General Accounting Office.

The Obama Administration wants to change that so that at least State and local HazMat and public health officials are aware of the chemical effects and formulations. MSDS sheets may also be required.

There are about 17,000 secret chemicals. 151 of these chemicals are produced in amounts larger than 1 million tons per year. 10 are specifically used in children's products, according to the EPA.

Chapter 5: Environmental Philosophy

Paying for Environmental Pollution of the Past

Who pays for environmental sins of the past effecting a few local towns versus the worldwide environmental sins of today affecting millions? Who should pay? Big Business? Big Government? You? Or no one? The Comprehensive Environmental Reclamation and Cleanup Liability Act (CERCLA), known in decades past as Superfund, now more commonly on the local news as HAZMAT, was in the Senate in the 2010 election year. Of course, prompted by BP's egregious negligence and crime. The Interior Department led by Ken Salazar ignored the government hacks and appointees in the agency overseeing off shore oil drilling.

Why should current profitable business pay for decades old environmental criminals? For that matter, why should the Federal government? Or even state government? Why not your town property, income and sales tax? It affects your water, your air, and your children. Why should I care in Connecticut about the polluted industrial states in the South and West?

CERCLA expired when the Newt Gingritch Contract On America was signed into office in 1994, rejecting the Liberal agenda of Clinton/Gore and the Intelligentsia. Corporate funds from CERCLA dried up years ago. For the last decade under both parties modest Federal funding went to clean up waste sites owned by old businesses large and small.

Should Big Business pay now? Why should the Federal government pay? At most it is a county, local, regional problem, like BP is to Eurasia. Local problems must be paid by local citizens, not the Federal government.

As the Climate Change Control movement struggles to refocus after the failed Copenhagen Conference in December 2009, the UNFCC and the whole movement is lost. Climate Change Control in the United States Senate is all but dead. This is good news. The Waxman Markey House bill passed in June of 2009 was an industry pork barrel. The Senate attempts to pass a Climate Change Control bill in 2010 were even more of an industry sell out. It may just take a Republican President in the next 10 years of the stature of Richard Nixon or Ronald Reagan to pass a tough Climate Change Control bill in Congress.

Amidst a continuing world wide sluggish economy and a jobless recovery the UNFCC is irrelevant as is the United States Congress. Climate Change Control advocates who continue to urge national and international legislative bodies to adopt mandatory, internationally binding Carbon cuts will wake up year after year to dismal disappointment and failure.

Even if Clean Energy is the new job growth engine world wide to re-start the economy and provide new jobs in every sector, it is time to de-couple the Climate from Energy. Climate equals abortion in the eyes of the die hard skeptics and the environmental fanatics. It should not be so political or polarizing, but it us.

Energy seems to be a euphemism for what is really ailing the world economy, reliance on 200 year old dirty Energy sources that claim hundreds of human lives every year around the world in mining, drilling, and explosive fire incidents. The cost of Conventional Energy is simply no longer worth it on a fully justified economic cost-benefit analysis.

In other words, politicians and the markets have externalized and hidden many deleterious society and property loss cost impacts from Conventional Energy. In addition to worker lives lost we all get the unpaid cost from Conventional Energy of air and water pollution and hazardous waste management for a cheap utility and gasoline price. Other costs are emergency preparedness and response to natural gas, propane, petroleum, coal and nuclear plant explosions, petroleum refinery and tanker explosions and fires, mining collapse disasters and shipping accidents and spills.

Wind and solar and to a degree electric cars have a much better total economic cost-benefit full value price than Conventional Energy of coal, petroleum and nuclear.

Renewable Energy is more expensive to electricity consumers now but this is distorted because we are hiding the total value cost from Conventional Energy.

The Carbon Tax Cap and Trade system will place a price on Carbon throughout the world in the next few years. It is already happening in half to two thirds of the world economy. The Clean Energy Carbon Zero economy of the next several decades is here.

Environmental Protection in Great Recession and Wars

Environmental protection was neglected amid job loss, war and renewed terror threats. Many Federal environmental protection acts have expired in the last two decades. Voluntary monitoring and reporting and lack of enforcement have removed environmental compliance from corporate budgeting, and are only used for public relations.

Environmental protection, the biggest government success story of this generation, is now viewed as obsolete and the realm of Intellectuals. The Middle Class dose not value public health protection from environmental law. Environmental law is now a state responsibility, with local government barely able to respond to the episodic HazMat incident.

Lives saved and jobs created from environmental protection at the national level are the last thing analyzed by EPA, Interior and OSHA due to funding cuts. State government does not have the manpower or resources to evaluate public and economic benefit of environmental protection.

Continued job loss and unemployment, war and renewed terror threats both domestically and internationally, mute our reliance on foreign oil and environmental protection.

It is unlikely Super Fund, Clean Water and Toxics Control will be re-authorized any time soon. Climate Change trendiness and scientific corruption and advocacy paints most environmentalists as extremist sympathetic to Greepeace and PETA, things the Middle Class find abhorrent and repugnant. Middle Class stabilization of family finances is more vital than their health. This is not a political issue but one of economic cost benefit, a mundane exercise the Middle class cares less for in Recession and War.

Do Voluntary Controls Work?

The history of industry self reporting and policing is a largely a one of dismal failure from the breakfast cereal industry to Wall Street to environmental polluters. All abuse and neglect a self policing system. The Obama Administration can not seem to move fast enough to plug all the holes in the nations regulatory and oversight control programs.

In an odd twist, industry, especially in Texas, has taken non-compliance with environmental law as the baseline for cost. So, enforcing existing environmental law gets the absurd result from the Congressional Budget Office that enforcement of air quality regulation in Texas will cause great job loss and economic harm. It does not seem like four years will be enough for Obama to undo the harm caused by Bush and Republicans in their pro-business agenda.

And worse, the Climate Change Control bill passed by the House and the one winding through the Senate are industry give a-ways, providing the majority of Carbon pollution permits to industry for free.

The Senate put a ceiling on a ton of Carbon at $28 per ton. Carbon trading experts think the floor is around $15 per ton and will stay flat through 2025. The average U.S. household creates one ton of Carbon a month, so there is your cost. The Senate bill gives you that $100 a year back in utility discounts from the paltry amount of credits auctioned off.

The UNFCC is looking to scrap mandatory international limits on Carbon emissions. The plan now appears to set voluntary goals of Carbon reduction and investment in Green Technology.

This is a step forward, but not a very big one. The Carbon trading experts say we will wake up in 2020 only to find we've run out of cheap agricultural and ranching offsets (paying them for many things they already do, like no till farming). In 2020 we will find that we have missed our Carbon reduction targets.

It is clear in the new international economy that regulation and control have taken the back seat to economic growth. The sad thing is that economic growth is coming with very little job creation. Hence the public stomach for government intervention in the markets to correct societal external costs is extremely low.

Massive job losses in Texas and Louisiana under Obama environmental initiatives are to be expected. Right at the time Obama and Congress were crafting legislation on Climate Change, the EPA is stepping up permitting and operating pollution controls on the petroleum and natural gas industry in Texas.

Bush era EPA officials used a voluntary compliance program and did not enforce the Title V Operating Permits and New Source Review rules of the Clean Air Act Amendments of 1990 in Texas, many Gulf States and elsewhere across the nation.

Texas feels it is being singled out in retribution as the home state of President Bush. Between Climate Change Control programs and Clean Air Act enforcement, the Congressional Budget Office predicts "major and substantial jobs losses and impacts on communities" in Texas, as a result of the Obama environmental push in Texas.

The House Climate Change Carbon Cap and Trade bill provides zero free allowances to the oil and natural gas industries dominating the Texas economy. While the Congressional Budget Office said there would be job growth from the Obama environmental initiatives, they are spread locally around the country, more evenly than in Texas and other Carbon based fuel industries in the Gulf of Mexico.

Texas is fighting back, but making no progress. After lavish riches under Bush, and fat contracts for Defense contractors, like Texas based Halliburton, the next generation of Texans will face great job and economic uncertainty.

Mandatory versus Voluntary Carbon Controls

Republican Alaskan Senator Murkowski has no impact on the regional Climate Change Control programs running in the West, East and surprisingly coal rich Midwest. While the South and a few small market state companies are not covered by voluntary programs, 75 percent of the U.S. economy and Carbon is controlled or will be by the regional groups. Only Texas is a big enough player to continue to try to violate the Clean Air Act.

Senator Scott Brown (R-MA) the upset victor by Independents in a special election in January 2010, met with President Obama to forge an agreement on Energy. Senator Brown dislikes government mandates and quotas on anything, especially something as vital to our daily life as Energy. Obama agrees, but says the market barriers for Energy Technology are huge, regardless of fuel source, newness or timelessness. Brown agrees.

It appears both are settled without further Federal action on Energy that the United States is relying on the traditional innovation of individual states and the private sector to solve Climate Change. The larger established Energy interests will exploit the intellectual resources of new Energy Technology innovators. This promises much for employment in the next few months and years ahead.

All agree, Energy is vexing problem, intractable and almost impossible to solve. All agree the nation must diversify fuel sources and power supplies. Yet the cost of the research in everything from clean coal to harmless wind turbines to low maintenance nuclear plants and disposal of waste are daunting.

Many goals are not achievable fast enough to solve many societal challenges. Both agree, and regret, the Federal government can not solve the problem without help from as many different resources, classes and nations as possible.

This is not just a United States problem. Every nation has Energy challenges, either creating or distributing Energy or making it safer, cleaner, and cheaper compared to the current supply.

Both agree current suppliers, like BP and Exxon, will be the suppliers in some form 100 and 200 years from now. Fossil fuel will likely remain the least cost option even with controls, some still in the lab.

The Brown-Obama bipartisan Clean Deal of 2010 combines modest government goals and finance with healthy investment from the private sector and research universities to create jobs now and in the days ahead. This will diversify the Energy supply source, while cleaning up older sources and lowering costs of newer ones.

The Energy bill, the re-birth and re-branded Democrat Climate Change Control - Cap and Trade bill is an industry give away. Or is it an environmental benefit?

It is neither.

It is a political bomb that will never pass, should never pass and best left to the states. Government mandates do not work. Sure, it helps, but as soon as the economy goes south or a different perspective is elected or we simply forget, regulation and compliance go by the way side.

What works then?

Self interest and consumer product switching to firms whose economics and sustainability match their style, budget and convenience is the cheapest and quickest way to Control Carbon.

Industry chafes at government control and regulation. But it hates lost market share due to poor design, operation, maintenance or health effects on the individual, wildlife and the environment more.

This is how the market should work.

Funding clean coal, off shore oil drilling, Clean Energy research and development and a host of other seriously non-Climate Change Control projects are the pork Obama and the Liberals are shoving down the country and trying to pick off Moderates in the Senate.

There are rumblings of withdrawal of support for expanded off shore oil drilling from Coastal Democrats. This is a minor issue. Industry will only economically drill off Alaska and the Gulf, not the East or West Coasts.

What about the coal state Moderates and resource intense states? How is Obama going to buy them off? He can not.

The economics of Climate Change Control at the Federal level for them is political suicide, even if the region already has a Cap and Trade voluntary state program starting to work. Forget about Texas.

The regional Carbon trading groups run the country, economy and Carbon market. We already have Carbon Control. Let EPA tackle backward states like Texas and the South.

Let it play out in the Courts. EPA will win. They already have. Climate Change Control is almost two years old and starting to work well. This did not kill the economy, because reductions in Carbon are not from control but a depressed U.S. economy. Climate Change Control is here.

I heard about digital technology replacing analog technology in college in 1980 with the advent of the CD. I've heard about a fossil fuel free economy since childhood when the Conservative Era spearheaded by President Nixon began and then culminated by President Reagan.

What does the Carbon transition have to do with the digital transition? Economics plain and simple Digital technology in 1980 was too expensive for the mass market and there were still technical challenges.

Now we have MP3 and Smart Phones. None created by government mandates, health, safety or security interests. Pleasure and mobility, ease of use and economics drove the digital transition. Carbon free Energy is no different and much more challenging to achieve.

On the eve of first national referendum elections on President Obama and the Neo- Liberal Era, states have created voluntary Carbon Control Cap and Trade programs in every region of the country and every economic sector expect the Gulf States and petroleum industry, and they are not far behind.

The signature of the three regional Carbon Control programs in a majority of the states is Clean Carbon Free Fuel specific power source generator supply contracts by 2020 to reduce Carbon by 20 percent.

Do voluntary programs work without mandated market diversification regulation? So far in the history of both tried alone either environmental regulation with government mandates or voluntary industry compliance with regulation is a failure. Neither approach is flawless. Both are to prone to abuse and fraud.

The government mandates from President George H.W. Bush for Alternative Transportation Fuel in the Clean Air Act of 1990 and the Energy Policy Act of 1991 both were relaxed by enforcement agencies EPA and DOE and never met.

Transportation sector imported foreign petroleum dependence is even stronger in 2010 than in 1990 despite war in oil rich regions of the world.

The voluntary industry compliance with virtually every government Police Power agency, created by President Nixon in 1970, in effect under the Republican Congress of Newt Gingritch from 1994 to 2006 failed miserably and now we teeter on the verge of Double Dip Great Recession.

So what works? States are good breeding grounds and policy labs for national Federal policy. The only Police Power the citizens seem willing to concede is law enforcement against crime and national defense.

The Police Powers of the Constitution enacted in dozens of Congressional Acts by both political parties since President Nixon created modern Big Government with EPA, FHWA, HHS, DOE, OSHA and smaller agencies are viewed commonly as central command and control towers without economic analysis by faceless unelected Federal employees.

At least this is what the Republican's leader Massachusetts Senator Scott Brown said during the worst of the Deepwater Gulf of Mexico Oil Spill. Senator Brown espoused strong support for BP, Big Oil and Big Business.

Has the Obama Liberal Progressive Change Revolution nascent two years displaced the Wild West mentality of Corporate Republicanism Government crime and corruption?

Wall Street and finance remain barely regulated. No change. Texas and the Gulf of Mexico states ignore Federal environmental, health and safety laws fighting in Court to succeed from Federal jurisdiction.

This is except for national defense and emergency response as in Katrina and now the BP Gulf Oil Catastrophe of 2010. Can Big Industry, Big Oil, Big Coal and Big Finance self regulate in lawlessness in 2011 and beyond?

You judge and vote.

You have the Contract On America Republican Square Deal, two Wars, the Global Recession and a few years of Progressive Liberal Change Management.

Who do you trust to give you the career you are trained for and your children educated in after January 15, 2011?

Do you want it to be Energy Technology and the Clean Deal? Or Dirty Coal and Dangerous Oil?

Which job pays better for the high school graduate in 2020 with least risk of fatality to the worker? .

Regulation Only Sometimes Work

The modern industry regulation is only 40 years old. It was established by President Nixon before Watergate. Nixon established EPA, DOT, DOE and OSHA. President Ronald Reagan greatly relaxed and de-regulated industry in the early 1980's, leading us out of Recession and vastly improving the efficiency of all transportation sectors.

The only regulation that has work for the past 40 years is environmental regulation from EPA and the Interior Department.

But even this has lapsed at time for years, not re-authorized, delayed, amended or not enforced. The surprising thing is that powerful industries that are the cornerstone of the U.S. economy, like Big Oil and Big Auto, have been successfully regulated more tightly over the past 40 years.

There will always be swings in the public, private and government interest in regulation, depending on the economy, safety and health threats. But on the whole regulation of industry works cost-effectively for the human lives saved and property damage prevented.

Command and control regulation is not favored by the public or politician. But every 20 or 30 years it is necessary to make progress on vexing public issues.

Market Based Incentive regulation is far more cost effective, easy to implement and more palatable to industry. When it is made voluntary it has proved ineffective at solving societal problems.

A command and control component is always a part of market based incentive programs. The most effective market based incentive is the Cap and Trade program established 20 years ago with the trading of sulfur and more recently nitrogen credits.

The price per ton came in far less costly than industry or even the EPA estimated. Real reductions in Acid Rain and Smog have been made because of these programs.

The Congress would be wise to pass Cap and Trade for Carbon. In an election year, with resurgent Republican power, this is unlikely until 2014 or 2015.

Energy Policy Less Political, More Individual

Energy policy is becoming less a political party issue and more of an individual candidate's issue. Sure, the major parties will still sensationalize, seizing on its personal appeal for media coverage. But Energy use is the world economic engine.

The world's Energy consumption is growing rapidly as the Emerging World develops modern lifestyles. This is very positive for the world economy and for all Energy types.

The coming several decades will see needs for every source of power generation the world can muster. From coal, oil and nuclear, to natural gas, propane, wind, solar, tidal, hydro and biomass, the world will need them all.

Concerns for externalities are now commonplace in Energy siting, development and construction. The world is properly pricing both the known and the unknown external societal, health and economic impacts of all types of Energy exploration, development, use and disposal.

Energy is so vital to our very life. Perhaps Energy is right at the top next after water, food and shelter. The world has progressed rapidly in the past half century under multiple political views to internalize the externalities, both known and potential, to attempt to reduce them from all types of human activity.

As the Congressional and State elections swing into gear, the parties will offer slogans of their solutions. The Right, usually provides merely lip service. The Left, usually mouths action but provides few resources. The Kennedy's opposed Cape Wind. So did Scott Brown. Governor Patrick and Obama praised Cape Wind.

The Obama Administration is funding modest new nuclear facilities and clean coal. Also there are more funds from President Obama for wind, solar and other Clean Energy through Federal grants to State and local government. Governor Patrick has astutely used free Federal dollars to vastly expand Clean Energy generation in Massachusetts. Several other states have also.

Pricing Carbon and environmental damage from all commodities, especially Energy, is wise. The free market will decide the outcome. For a very long time coal, petroleum and natural gas will remain the cornerstones of the world economy.

Yet the world economy is diversifying its Energy sources and locations. Many reasons dictate this reality. Intangible military costs and political outcomes force the requirement for Energy localization and decentralization for the next several decades. Energy, food, defense and monetary policy always glue the world economy together.

More localization, regionalization, diversification and reduced externalities will strengthen the economy, provide local jobs, and meet growing Energy needs while reducing the known and mostly unknown effects of Energy consumption.

Environmental Law Enforcement Used Wisely

I remember as a young regional planner in the environmental air division in Massachusetts years ago receiving call a call from a self reporting polluter.

It was a serious public health hazard I am in planning not enforcement. I had no experience in enforcement but from my education in school and conversations with co-workers.

I asked the caller if they were planning on hiring a licensed contractor to clean up the hazard. "Are you going to fine me?" the voice came over the line.

"I am more interested in the positive result of solving the problem than exacting a fine civil or criminal against you. Please clean this up as quickly as you can with a licensed contractor", I replied.

I told my manager who reprimanded me but I was not discharged. The end result was unknown but I had no option available to me on the public information hot line for asbestos calls.

I view my job as a legislative and policy analyst who sometimes writes or assists in the forging of environmental law and policy as a matter of intent and spirit, not precision and exactitude of the letter of the law.

There is law, regulation and policy.

Law sets the framework for regulation which details the parameters to comply with the intent of the law.

Policy is the administrative implementation of law and regulation. Policy changes with every Administration at both the State and Federal level.

Policy responds to public perception, the economy and the goals of the elected leader of the State or Nation.

Individuals may have their political and environmental agenda, but as a civil servant, directly reporting to elected and appointed officials, or as a contractor hired by an agency to assist them, I follow my management team's instructions of the current policies in place.

My role is not so much to follow the letter of the law but to follow the spirit and intent of the law. Most people in enforcement are more interested in the citizenry and industry following and complying with the law than in exacting civil or criminal liability.

Violations will and do occur. The EPA and state regulators could fine industry out of existence technically.

It is not politically, economically or socially responsible. I have never seen a nefarious enforcement agency exact an unnecessary fine against a violator of environmental law.

True, egregious examples of intentional malice and forethought in harming the public exact the most disgust from the public and receive the severest public punishment. This applies to the headline news serial killer and the midnight toxic polluter of your water supply.

But the cases of egregious and flagrant environmental violation are rare. It is much less likely in a world of increasing New Media and Web 2.0 writers to report bad corporate citizens.

This is especially true of errors unintentional that harm the public harm the corporate image and erode market share.

Shareholder movements to change corporate actions lead by groups like Ceres of Cambridge, MA work with industry to voluntarily create sustainable, less polluting products and services. This is the model of the future.

The political will to crack down on coal, oil and other power generators simply does not exist because the devil you know is better than the devil you do not know.

This may sound corrupt or jaded. It is pragmatic. We need and will continue to use both as a nation and as a planet, coal, oil and nuclear power for the next several hundred years.

We know there are problems with all of these Energy sources. A new generation of environmental activist is pushing for zero discharge permits into waterways and zero emissions into the air ways. This is lauded. The automobile is the closest machine to achieving a zero emission impact into the air thanks to decades of dedicated automobile engineers in the United States, Japan, Germany, South Korea and now China. It is easy and simplistic to paint industry with the brush of the bad boy.

True, many coal plants and petroleum refineries have skirted air and water regulations for years. But as each week passes more and more sources of pollution that bent the rules see it is a better public relations tool to drop the paper charade and install the pollution abatement and process improvement controls.

The desire for zero pollutants is not pragmatic today but a laudable goal, like Greenhouse Gas reduction and Renewable Energy targets for 2020, 2050, 2075, and 2100.

In election years it is now common place to rail against regulation of any type. It is an easy sound bite. It makes the government look bad, out of touch and evil.

We need all the regulation we can we can get. Rules do not run the economy or you, they guide you, like your doctor or your workout trainer.

Enforcement is where the rules could but rarely do bite the economy. True, complying with regulations costs money. But there are hundreds of economists and consultants like myself who can find least cost compliance options.
Often time solutions involve new technologies that create more jobs as a multiplier than the temporary Management Consultants hired to devise the plan to comply with the regulation.

Enforcement, or the Stick, is seldom used unwisely by government. The Carrot is always the best way to get industry to comply with the law.

The Carrot approach to regulatory enforcement also paves the way for the revolving door between industry and government.

This is citizen government after all. It injects new ideas and perspectives into both government and industry after neutrality periods and conflict of interest laws are met.

Regulators and industry are the heads and tails of the same coin that run the economy and protects public health and property. Neither the head nor the tail can withstand bad publicity, slashed funding, or lost market share due to over excessive breach of influence.

Dualistic Air Quality and Carbon Control

Essentially, I am a life long cultural anthropology student. The two most influential researchers in my life work are the ones from junior high school. I read an anthology of a few cultural anthropology researchers.
Bronislaw Malinowski, a German political dissenter in the Trobriand Islands during the Nazi Era and Margaret Mead, a Liberal United States Feminist, who studied several major cultural threads, were early influences.

Many anthropology majors express their studies in Urban and Regional Planning, as I have. But my studies at Phillips Exeter Academy after studying Latin, History of the U.S., Europe and the Holy Roman Empire through Post World War II America went on in college to Non-Western studies, every topic in Anthropology.

I quickly learned most major problems in both the Western and Non-Western World were centered on two major issues: Urbanization and Environmental Protection.

As a Liberal Arts BA and MA, I barely comprehended math and science. I remember freshmen year studying the Edison Electric Institute scientific reports on the very complex chemistry, engineering, fate and transport and meteorology of Acidic Precipitation and Deposition.

Senior year I would work again on Acid Rain. This time I studied, wrote and spoke of the politics of the Clean Air Act, its expiration and need for re-authorization in Congress. I was unaware that my preparation would become a focus of my professional work, not even as a goal or dream.

Signed by President George H.W. Bush on November 15, 1990, the Clean Air Act, as Amended in 1990 was the first major environmental protection law passed by Congress and signed by the President with no "Sun-Set" or expiration date, like every other major environmental protection act since 1970.

I got a job on the Clean Air Act of 1990 in Massachusetts state government. Then I moved to Maryland for a job with a consulting firm with key contracts in Pennsylvania and Missouri, and the United States Environmental Protection Agency and U.S. Federal Highway Administration.

Political and controversial, many of the major air quality control programs I helped plan and implement 20 years ago are now operations programs barely noticed by the public.

Today it is Climate Change Control legislation and Energy policy that are again under political legislative and funding debate in the Senate.

As we go forward with Carbon Control programs, already functioning in most of the United States through voluntary state Regional Climate Change Initiatives, I suggest air quality control planners not forget Ground Level Ozone and soot dust from fuels.

Many Carbon friendly fuels and options in study, evaluation, deployment and assessment are not heart-lung friendly options despite lower Carbon content. Carbon Control must not sacrifice the more mundane problem of summer Ozone weeks in the U.S. for the majority of the United States population which breathes and walks laboriously outside on Ozone pollution days, even when the state says the Air Quality is Moderately Healthy.

In planning for Carbon Control options air quality control planners need keep in mind the Ozone suffers of asthma, the elderly, outdoor workers, children, cancer survivors and those with heart or lung impairment.

A majority of the nation's citizens need protection from the old, common and known pollution problems, while we engage in political debate and program assessment of Low Carbon Control options for the future.

Fuel Neutrality and Government Policy

The essential fuel policy of government analysts for the past 50 years is one of neutrality. Let the market find the most cost-effective answer as long as the government standards are met reasonably in time.

This has changed in the states with voluntary Carbon Controls. States are allowed by Congress to enact in legislative acts more stringent than the Congress' and demand target goals for the environment and safety than those that are required by Federal law, legislative or regulatory.

23 states have a strict requirements for power sources to buy large amounts of Renewable Energy like wind or solar by 2020.

Congress would be best to not to act on the American Power Act or Energy reform, since it is compromise with the fossil fuel and nuclear firms that run the economy to dangerous safety, security, mobility, and defense costs of the nation, while a few get rich.

The country ships billions of dollars out of the domestic economy every month for transportation imported petroleum to dangerous foreign countries. Fossil fuel is the New Terror.

Favoring wind and solar, even though they do not meet the economist cost-benefit, market efficiency test is good policy, given the larger costs to society of fossil fuel.

Massachusetts residents favor Carbon Control and Cape Wind, despite impact assessments claiming high costs in several evaluation areas, for a virgin Energy Technology power source.

This is bi-partisan policy set at the state level by both parties in a majority of states in each chamber of the legislature and signed by dozens of Governors in the past few years. This is not politics. This is the Clean Deal of Energy Technology.

Marginal Utility and Economics

Marginal utility and economic efficiency are economics class basic concepts of increased productivity from mass consumer commodities in small increments and low cost to create very large aggregate savings to the producer.

The best way to understand is called the parable of the "Tragedy of the Commons". In Colonial Boston settlers took their cattle to the nearest field to graze on the grass and laze in the summer sun all day.

In low density populations this common good and shared resource is enjoyed by all property owners. When the population density increases beyond a carrying capacity to support the large number of cattle grazing on the green, it is over harvested.

Without regulation and control, the public resource can not sustain regeneration and is depleted beyond use and recovery.

The Tragedy of the Commons is most pronounced in the 5,000 year old earliest Western cultures of the Greeks and Lebanese. The now barren rock surfaces of both Greece and Lebanon look ugly today.

In early civilization both countries had lush, green, bountiful hillsides with an abundance of valuable forests. Unfortunately, neither culture self regulated and managed their natural resources. Both were over utilized and for perpetuity are barren, benefitting no on in society.

This is a major theme in the United States Interior Department and Environmental Protection Agency.

Off of the coast of Boston and New England, fishermen and lobstermen have over harvested Great George's Banks, one of the world's most plentiful fishing breeding and commercial market resources, for decades.

The United States Federal Government through numerous different agencies has sporadically placed moratoriums of fishing to allow the depleted fish stocks to recover to serve the market of fish lover in America and elsewhere.

Public natural resource and market workers never self regulated and abuse the bounty of the precious commodity that supports their community and families.

The government steps in to protect the resource from greedy, self-serving misuse and poor industry management with regulation. Without regulation the natural resource crashes like the forest of Greece and old Boston Common. The regulation will recover and heal the Great George's Banks fishery for generations of economic utility for the markets in future centuries.

Governments around the world are confronting Climate Change Control today. Depending on which country and economy analyzed, different local and regional options to manage Energy use and reduce the Carbon foot print of the economy are in place or in the planning phase.

As the United States Senate debates various legislative Energy diversification and Carbon Control options, the over use of forests in Greece are best remembered for the future of the American way of life.

Advising Elected and Appointed Officials

I have spent most of my career assessing emissions trading. It is something developed during the Carter/Reagan transition to preserve environmental law and regulation, human life and property while growing the economy.

The law of averages or "bubbling", as President Reagan would explain during his two terms in office, simply captures and controls the biggest and most dangerous air pollution source first. Then it brings in new, cheaper, efficient technology to re-invest in the economy, job growth and management of the older less competitive industries.

President Reagan knew many aspects of the old Industrial economy were not the future economy against Japan and Europe.

It became the standard government agency rule making policy. Influenced heavily by cost and benefit risk assessments, it blends many different types of professional backgrounds into a coherent national environmental policy, based on human health risk, property loss and economic impact.

There are times when one of the variables will derive the answer of the particular policy or regulation. Some politicians and lay people, usually directly disproportionately affected by the certain government action attempt to brand all government environmental policy as Marxist or Fascist.

This is fear mongering and ignorance. It must be ignored. It is not how a majority of the scientific body and policy developers create environmental protection law and regulation for the Congress and Legislatures to enact.

Science, Uncertainty and Lobbying

The Minnesota Legislature, like many others with large urban counties and air pollution, had to vote on several things about the Clean Air Act as amended in 1990.

A legislator's comment was, how can the elected representatives make a simple yes or no vote on a very complex issue that involves, pretty much a dozen or so economic and technology issues? These rules directly cost the public now with the future benefit of improved public health and reduced property damage unknown.

I did not understand his confusion. It was clear what the answer and vote should be, based on science and technology. Now as a new Clean Air Act issue is in the political spotlight, Climate Change Control Carbon Cap and Trade and Clean Energy, I have a better understanding of his confusion.

Science, like any community, has a diversity of opinion, like every community in the world.

Consensus and clarity are rare and hard to replicate.

Einstein never learned this. The founder of Absolutism and prediction with clarity rejected the earth shaking Heisenberg Theory of Uncertainty and Quantum Mechanics.

Uncertainty Theory and Quantum Mechanics are the economic driver that transitioned the polluting industrial world economy into new, life improving, Clean Technology (for the most part) we continue to explore.

So what is the Politician to vote for? He is bound by polls and elections representing his constituency. The Clean Air Act of 1990 had 18 month Federal deadlines. If the deadlines were unmet it meant withholding of new highway funds, a huge job and economy transformer at the local level.

There was much success in some states with this deadline, Pennsylvania, Missouri, Maryland and California, where I consulted, not so in Massachusetts, where I also worked.

You can not put a clock and penalties on the Climate. It is doing what nature does and we have not, will not, ever reach unanimity of consensus in the scientific literature on it, or any science topic. Always new research results. That is what scientists and engineers do best.

Not lobbying.

Yet Congress charged the National Academy of Science, a peer run science body that sometimes advises the White House, to answer some questions on Climate Change Control Cap and Trade legislation in 2008. In 20010 the Academy strongly answered: Take Legislative action on Carbon Control and Cap and Trade based on a review of the science.

The Congress required the National Academy of Science to do it in law. The Academy met this law's requirement. Act now. Legislate Climate Change Control law.

I understand the different views about controversial legislation that affects jobs, industry and the average person. The latest scuffles surrounding Climate Change Control government legislation and regulation at the national level are understandable to me.

I am not a scientist. I can not prove or disprove Climate Change. The environment of Climate Change Control policy is polarized and personal. If you are a doubter or feel we are wasting time by taking so long, you go on to writers you believe.

As a contractor, I understand limited government funding in a competing world of public interest goals. Often times these policy issues become deeply divisive and political. It should not be this way.

Throughout my reading and essays, I have only one belief, other than God. I agree to disagree.
I do not discard those on either side of any issue, of their right to form and hold their point of view because of my belief. I may not follow their traditions or be in their groups, but I accept and appreciate the difference.

I some times pen essays on Climate Change Control. My experience in Clean Air guides me to understand that creating common ground briefly for the 35 seconds you may land on this page is necessary. I have written dozens of one page memos to leaders, educating them, like here, on something they must chose an outcome for affecting those involved.

Climate Change Control has already won legal standing by the Supreme Court. Some in Congress want to strip the United States Environmental Protection Agency of funding and delegation to regulate Climate Change Control.

The National Association of American Manufacture and the United States Chamber of Commerce have lost key industries as members due to their lobbying and advertising against Climate Change Control. The two lobbies have pulled back from there offensive, while silently opposing the bill in private meetings with Senators.

Climate Change Control should not be a partisan issue, yet it is. Different regions of the country have different core economic engines. In the Midwest the base job provider is coal, and in the South, oil. It is only in self interest that these Senators vote against the American Power Act. Their supporters are against the bill.
Fundraising and elections are more important than a vote for something they misperceive as a threat to jobs in the home state.

Los Angeles, California has the most air quality controlled and regulated regional import-export economy in the world. Since Los Angeles and California started air quality regulation in the 1950's, the economy has continued to grow and boom.

The Los Angeles and Orange County economy has been adapting to new air quality regulation, inventing new technology and creating new jobs for all parts of the regional economy.

Senators in oil and coal states would be wise for briefings on the history of the relationship between air quality legislation, regulation and increased exports in the Los Angles international economy before passionately voting against the American Power Act.

Obama spoke in support of the American Power Act in 2010. President Obama also spoke against continued tax breaks and subsidies to all fossil fuel companies.

Obama said, "Inherent risks to drilling 4 miles beneath the surface of the earth", in order to sate the nation's OPEC imported oil dependence causes lost domestic Energy Technology jobs.

Energy Technology promotes research, development, production, distribution and local sale of Clean Energy such as biomass, tidal, wind, hydro and solar.

Obama said, "We consume more 20 percent of the world's oil, but have less than 2 percent of the world's reserves. Without major change in our Energy policy, our dependence on oil means that we will continue to send billions of dollars of our hard earned wealth to other countries every month including countries in dangerous and unstable regions".

Obama also spoke in early 2010 of ending all fossil fuel tax breaks and subsidies, including the ethanol tax break, decades old, to corn producing states. This is a much more political and highly charged local and state politics issue. Threatening to cut off the profit boosting Federal aid to fossil fuels is not new, but highly contentious

Obama wanted to use the saved fossil fuel funds, giving it to Energy users and Energy Technology firms to offset any unintended consequence of the American Power Act.

The future of Carbon Control is certain. It is here and it is working worldwide. Point Carbon, an international Carbon broker, pegs market price through 2020 at around $15 per ton.

The other established air quality commodities sulfur and nitrogen trade for around $750 per ton.

Carbon is cheap, plentiful and easier than thought to switch away from. Yet vested interests will maintain the grip of Carbon fossil fuels for hundreds of years in the future. Fossil fuel is economically inexpensive, has a well built and maintained support structure from exploration and recovery to distribution, worldwide.

Non-Carbon sources are relatively new and are expensive. Government mandates and Energy Technology varies from region to region and nation to nation.

Clearly, a hot button political, personal and pretty much any reason you dream, for or against Carbon Control and non-fossil fuel.

But we know Carbon based fuels are finite, and becoming more expensive to harvest. The Gulf of Mexico oil tragedy is unfairly pinning BP and the oil industry with the sin of the car, personal mobility and economic choice. In a few months it will pass from the glare of the masses and politicians.

Investors in non-Carbon Energy are few. If you invest in non-Carbon research firms, treat it like Vegas money. The odds are way stacked against its market penetration and the return in your life time in any value is limited.

Free Market Governments can not create non-fossil Energy industries. China and Europe are though, and will dominate non-fossil fuel technology and deployment markets, as they do today, for the foreseeable future.

Non-Carbon Energy is just too complex for consensus and meaningful bi-partisan policy sustained over decades to come, in most nations, to create a self sustaining non-Carbon Energy industry.

It is happening, just slower than the more pleasurable digital industry that went from the lab to the masses within my adult lifetime.

Non-Carbon fuel will not be prevalent as a source in many communities during my lifetime, either for transport or power.

But we try.

Many have, and will continue, to study, research, and apply ways to make non-Carbon Energy available to the masses.

Many will lose their money, careers and family for a dream still years away.

These blazers are knowledgeable and passionate. They will prevail eventually. When, we do not know.

Even a Manhattan Project style or push for the Moon, as the United Sates has achieved quickly before, would not solve this hexing security, safety and mobility handicap.

But we try.

We falter.

Better try, and take two steps forward on non-Carbon fuels, and once in a while a step back.

Progress on Carbon Control is here, it is real, and it's march is clear, regardless of how you feel.

Chapter 6: Climate Change Control Actions at the World Level

Climate Change Control Action in China

U.S. and China to this day remain far apart on Climate Change. China's refusal to cut Carbon emissions could derail the U.S. Congress from passing a Climate Change bill and signing the any successor to the Kyoto Protocol. China wants the U.S. to cut Carbon emissions by 40 percent by 2020, a goal even environmentalists say is unrealistic. Meanwhile, China maintains it will not make Carbon emission reductions at all. China and the U.S. emit half the world's Carbon emissions. Congressional delegations do hold Climate Change talks with top Chinese officials still. Members of the U.S. delegation before the Copenhagen meeting in November 2010 said they remained hopeful that China would eventually be brought on board to make Carbon reductions.

China could be come persuaded to make Carbon cuts since many of the Pacific Islands that face obliteration from rising sea levels might seek refugee status in China. Refugees from rising seas are a new problem added to the long list of global impacts from Climate Change.

Palau has introduced a resolution to the UN to make refugees from Climate Change a Security Council issue. Largely because the resolution requires the UN to take no action it is supported by the U.S., China and Russia.

China has been closing older, less efficient coal power plants throughout 2010. China adopted strict fuel economy standards for cars and trucks also. But China's growing economy is completely dependent on coal fired utilities. Until coal technology is cleaned up and sequestration and injection of Carbon Dioxide back into the earth, China will remain the world's largest producer of Carbon.

Many are optimistic China is headed in the right direction. Trade sanctions from the international community could eventually force China to join the Climate Change reduction community. But China as the world's largest bank could cut back on its lending in a retaliatory manner. It is unclear if the U.S. Congress will pass a Climate Change bill without participation from China and India.

Congressional leaders believed in 2009 that if Obama can lead the U.S. to craft a Carbon reduction bill, China and India would follow suit. Clearly the market for U.S. Clean Technology in the export market to China and India is huge. Hopefully, the U.S. will lead the way on Climate Change.

China Climate Change strategy is working. Slow economic growth during the Great Recession has reduced Carbon emissions around the world. No where is this more clear than China.

The International Energy Agency has reduced 2020 Carbon level projections by 5 percent. China is slowing its increase in Carbon emissions much more quickly than anticipated due to its intensive funding of wind, nuclear power and Energy Efficiency.

The Great Recession decreased Carbon by 3 percent in 2009 and 2010, the biggest decline in 45 years. Average annual Carbon growth in the past 10 years is about 3 percent.

The Agency aims to restrain temperatures from increasing more than 3.6 degrees Fahrenheit and decrease Carbon 23 percent by 2030. The cost is $10 trillion between 2010 and 2030. Lower Energy bills offset the investment. Each year of inaction increases necessary investment by $500 billion.

China is the world's largest Carbon producer but the U.S. is still biggest per capita Carbon producer, a more meaningful measure to reduce. If China can attain its Carbon reduction goals it would be at the front of worldwide efforts to slow Climate Change, according to thee Chief Economist of the International Energy Agency. China wants the U.S to cut Carbon by 40 percent by 2020 over 1990 levels. The U.S. has no Carbon policy now.

China is the de facto leader in global Clean Technology. Even though China is the world's largest emitter of Carbon and is almost exclusively dependent on dirty coal, it is the global leader in Clean Technology manufacturing. While investment in the U.S. in Clean Technology plummeted 85 percent in the last three years it continued to rise in China.

President Obama has said that the country that leads the way in Clean job creation in the 21st Century will lead the global economy. The U.S. sorely lags in Clean job investment and job growth.

Stimulus funding favored Clean Technology with about $100 billion going directly or indirectly to bolster Clean Technology. China is seen as doing more than the U.S. when it comes to investing in Clean Technology and jobs.

The new fuel economy standards of 35 mpg and the Carbon cap on transportation fuel at 250 grams per mile gave Detroit a clear roadmap as they emerged from financial havoc. But the White House has deemed the GM electric car the Volt to expensive to put on the road in the U.S.

Meanwhile, China is gearing up to become the world leader in electric battery and car technology with hefty government and Western investment. Even Warren Buffet is invested in China's electric car industry.

When it comes to wind turbines, two of the three world's leaders in wind turbine production are based in China. Even as China exports more of it Clean Technology overseas, it too is greening up its power supply.

Offshore Wind Turbines received a big boost from Obama when the Interior Department issued clear rules for the guidelines for constructing offshore Wind Turbine farms. There are no offshore wind farms in the U.S. but the path has been cleared for Cape Wind off of Nantucket Sound.

The U.S. has a long way to go to catch up in the Clean Technology race. Clearly stronger investment in wind and solar is needed from both the public and private sector.

However, traditional technologies that have gotten cleaner are fighting hard for the Renewable and Clean Energy status. Old sources like hydroelectric, new sources like land fill waste to natural gas and even coal waste industries are lobbying hard for billions in tax dollars and credits. Many believe this government subsidy belongs to new, under invested technologies like wind and solar.

The private investment markets need to step up their investment in Clean Technology. Right now, because of the relatively low cost of a barrel of petroleum ($61 in 2010), investors are not even investing in new petroleum production.

Yet, OPEC recently said to get ready for $200 per barrel of gasoline as early as 2011, as demand rises as the world economy recovers and production is not expanded.

Are we going to get flat footed again on high petroleum prices? Hopefully not, but this is the only market measure that will restart the investment in Clean Technology.

Unfortunately, the rewards of Clean and Renewable Technology are years away and the cost is now. We will have to see where we in are in the fall of 2012, of course, an election year.

There is no question China is the number one economy in the world. They emerged first from the Great Recession because they control their banking system unlike the West.

The Chinese and Russians do not have the problem with lax banking regulation. Their economies are much more stable and vibrant than the West. This is distinctively more so in China and much less in Russia. The Chinese people mostly like their government unlike most Western nation's citizens.

China, saturating the Western markets and government banking money supply, is now developing the rest of the world quicker than several generations of failed IMF and World Bank projects under the UN.

How did this happen that the Free Market failed?

Capitalism is alive and thriving world wide. The Chinese and Russians, while still Communists in name and theory, are practical business people, adapted to Capitalism.

The West failed by letting the illusory derivative market, bigger than all tangible assets, ruin the Western economies and governments. The result was casting hundreds of small companies out of business and millions out of a job and a home.

You are surprised the Chinese and Russian models are based on Mein Kampf, Adolf Hitler, and the Founding Father of the U.S. Hamilton's Federalist Papers.

Nazi Germany, Fascist Italy and the Empire of Japan all had state run Capitalist economies. The Totalitarians deadly genocide forced their extinction by the West. China and Russia are still repressive. Hopefully this will change like the other major state run economy in the world South Africa.

The U.S. will never adopt a state run economy, despite those who call Obama a Socialist. What the U.S. and Western governments must do now is bring in the unregulated derivative and banking system to a level matching China and Russia.

This is highly unlikely. Unless the West regulates the economy more, curtails oil dependence on the political and religious fragility of OPEC, the West will continue to decline compared to the in-name-only, repressive Communists.

The New York Times ran a lengthy discussion on China about its massive urbanization, Social Democracy and modernization of rural peasants. The peasant's small lot farms all over the China countryside are confiscated by municipal and provincial governments.

The government aims to increase food supply by creating an agglomeration economy of scale. The Chinese government hopes this will create efficient and higher yield food crops. The agricultural production and output should increase by creating industrial farms.

China's central, local and regional governments are granting the mostly non-literate peasants from the country side a greater Social Democracy voice.

The rural Chinese peasants now living in the cities are demanding modern conveniences such as factory floor air conditioning in manufacturing plants. The workers request to adjust downward to 70 degrees from the Federal Government required 90 degrees was granted.

The former peasants, now workers, want and are getting modern urban apartments. The units have efficient electricity and lighting. The apartments have efficient stoves and ovens. The Chinese people are driving larger more powerful low emission and high fuel efficiency cars.

The modern Chinese urban amenities requested by a large segment of China's 1.3 billion people are to simply join the modern Digital Age of civilized society similar to 1950's America.

The leading personal, social and economic freedom created by the U.S. government in the emerging Digital Age of Post World War II America was highway expansion and suburbanization.

This lead to the Eisenhower Plan requiring massive Federal highway funding paid for through the driver gas tax, of the Interstate Highway System. The Interstate Highway System was ostensibly for interstate long haul freight trucking commerce and National Defense in case of invasion from the Soviet Union.

China of the 21st Century, an urbanizing nation, faces many of the same challenges of the 20th Century U.S. suburbanization.

China is facing a growing mobile culture of expanding modern conveniences in the home, greater social mobility, individual economic growth and success opportunities.

China is meeting demands of its new urban residents in the same manner the U.S. met urban population Post World War II demand to escape the crowded, dirty urban cities. Suburbanization in U.S. led to the two acre large lot subdivisions in the nearby outer ring suburbs made accessible by the highway and transportation expansion in every town and county in the U.S.

Since the dawn of the Mobile Digital Age, Federal-State cost sharing through higher taxes to grant the wishes of a higher standard, civilized lifestyle have been chosen.

Both China and the U.S. today face anti-establishment Federal power protests across the lands demanding more local self governance and determination with less Federal restrictions on local solutions.

Carbon Control faces similar challenges today in China and the U.S. China, measuring Carbon on a unit of economic productivity, aims to simply offset Carbon growth between 2005 and 2020. China believes this will help it remain a Zero Carbon growth economy.

The U.S., measuring Carbon in the short ton of 2,000 pounds, is debating a threshold reduction of 20 percent of Carbon by 2020 over 2005.

This is the benchmark start year when industrial manufacturing and near full employment of the booming economy led to high Carbon yield. This level of carbon output will not be matched again for several more years as many plants simply shut down during the Great Recession. Consumers switched from luxury purchases of marginal utility to staples. Much of the U.S. began saving as the unemployment level reaches its highest level since 1989.

China is pursuing a Carbon Control strategy of no net growth while modernizing and urbanizing. The mature U.S. economy is primarily pursuing a Corporate State central command and control. This discount and reduction of the external, societal cost of Carbon into the Free Market Open Trading system allows Wall Street to fully price the entire life cycle cost of Carbon content in the U.S. economy.

China and the U.S. are both committing to control Carbon pollution to stabilize unpredictable weather patterns and exacerbated swings. Many researchers believe weather changes are due to excess Carbon pollution in the complex saturated ocean water and atmospheric gases water vapor cycle barely understood by meteorologists.

The dual approach will succeed in the New Technology Energy Age even as cultural resistance in political protests against rapid government policy change continues. Cultures and civilizations have long struggled to adapt to the realities of the problems created by advanced technology change and the classic demographic transition.

Climate Change Control Action in India

India paradox, going Green with no Carbon limits. Secretary of State Hillary Clinton was in India in 2009 before Copenhagen for high level broad talks with high level Indian government officials. She toured an Energy efficient office building outside New Delhi. The windows fill rooms with sunlight but keep the heat out. It was then that top Indian government officials expressed frustration over the U.S. pressure on India to make hard and fast Carbon emission reduction targets.

India Environment and Forests Minister, Jairim Ramesh said there was "no case" for the West to force India to reduce Carbon emissions when India already has one of the lowest per capita Carbon emissions in the entire world.

"If this pressure is not enough, we also face the threat of Carbon tariffs on our exports to other countries such as yours," Mr. Ramesh said.

The House in the Waxman/Markey bill passed in June 2009 would impose trade restrictions on countries that refuse to make mandatory Carbon reductions. Obama opposes such trade sanctions, saying they are illegal.

The meeting showed the stark differences between Developed and Developing Countries as Obama and Europe tried to forge consensus on global Climate Change leading up to the Kyoto Protocol successor in Copenhagen in December 2009.

Mrs. Clinton said the United States has no intention of making India give up its economic growth. "No one wants in any way, to stall or undermine economic growth that is necessary to lift millions of people out of poverty. The U.S. does not, and will not, do anything that would limit India's economic growth," she said.

"We are not in a position to take over legally binding emission reduction targets," Mr. Ramesh said at a news conference. "That does not mean we are oblivious to our responsibilities."

China also opposes Carbon limits. Both India and China say their economic growth must not be curtailed when the West did not encounter these limits during its industrialization.

Mr. Ramesh did soften his tough talk with the hope of "Green Technology" cooperation with the U.S. He proposes setting up Indian - American centers to develop Solar Energy and Biomass technologies and to study the long term effects of Greenhouse Gases. "It is possible for us to narrow our position'" Mr. Ramesh concluded, leading up to Copenhagen.

Green Technology is burgeoning in India and China as governments see ways to reduce Energy costs while growing their economies through exports of Green Technology.

Still the East and the West face the same hurdles, getting off of the hydrocarbon drug and dependence and finding a way to fund nascent, emerging, new, high technology "Green Technology".

Perhaps with cooperation both sides will advance some kind of political protocol in the future.

India seeks new path on Climate Change Control. India has softened its harsh voice on resisting Carbon limits. The change is brought about by India Prime Minister Manmohan Singh who is nudging India into a more international pose on topics ranging from trade to Climate Change Control.

Mr. Singh says India faces multiple complex stumbling blocks that are all interrelated. These include everything from domestic issues like pollution, Energy security and national security. All are tied into Climate Change Control.

Experts say India's previous stand did not work. India is seen as an unwilling laggard on Climate Change Control. China meanwhile, India's top development ally, has come off as more active on Climate Change Control, despite resisting legal limits on Carbon emissions.

"We can not compromise our basic national position on protecting our prospects for growth, but we can see things that can be done," said Nitin Desai, one of the Prime Minister's special Climate Change advisors. "The signal that I get is that India is not going to be a spoiler. If a reasonable deal can be worked out, they will be there."

Experts are looking at alternatives to mandatory rules on Carbon limits, like those of the failing Kyoto Protocol. As Copenhagen approached in 2009 planners were considering providing credit for domestic programs, like the one India is considering.

India's ties to China on development make it seem as though the two are equal emitters of Carbon. China is the world's biggest Carbon emitter. India emits a fifth of the Carbon emissions of China, measured both in total and per capita. China comprises 23 percent of all worldwide Carbon emissions while India contributes just 5 percent.

Some Energy experts who research India's Climate Change Control policies believes the developing nations will break up as India distinguishes itself from nations like Mexico, Brazil and China that have more mature economies and greater Carbon emissions. India is a very different kind of country than the top Emerging Countries in the world.

Mr. Ramesh, India's Environmental Minister, said China conveys a message that they are doing a great deal on Climate Change. Mr. Ramesh said India had two basic requirements for Copenhagen. Industrialized countries must consent to steep cuts in Carbon by 2020 and provide monetary and technical help to the Emerging World. India is still against mandatory Carbon limits.

Mr. Ramesh said despite all the bad press about India on Climate Change Control the Environmental Ministry is introducing legislation to the Indian Parliament to strengthen fuel efficiency standards, create voluntary targets to improve Energy Efficiency and promote solar power and implement clean coal technology among its utilizes.

At the U.N. in late 2009 Mr. Ramesh highlighted the basic needs of India like increased forestation cover, lengthening a treaty on investment in Green Technology and more technological help.

For a long time India defined its stance on Climate Change Control in geopolitical economic social justice and a national rights issues. India claimed the industrialized nations created Climate Change. It stated that the majority of response should come from the West.

Historically India has said legal limits on Carbon would stunt India's economic development. This debate is still raging in India as part of the domestic debate on Climate Change Control. Mr. Singh was rebuked by members of Parliament after consenting in 2009 in Italy to prevent worldwide temperatures from increasing more than two degrees Fahrenheit over 2005.

Mr. Ramesh said its his responsibility to create a new domestic political partnership on how India can fruitfully work on Climate Change Control without jeopardizing India's economy. "Without a solid domestic consensus, or even domestic constituency, we can not even think about engaging internationally. And this is also true of the United States", he added. "It's true of all democracies."

Climate Change Control Action in Brazil

Brazil, a non-signatory to the non-binding Copenhagen Accord, signed an agreement in 2010 to manage Brazilian forestry. Secretary of State Clinton was in Brasilia in 2010 to officially join the Brazil partnership.

Brazil and almost every other country including the U.S. are guarded against foreign intrusion on vital domestic economic matters. Forest degradation accounts for 15 percent to 20 percent of annual Greenhouse Gas emissions worldwide. Indonesia and Brazil are the 3rd and 4th largest Greenhouse Gas polluters, behind nascent China and rusty U.S.

It is impossible to put into numbers how much Carbon is saved from proper economic stimulus and Forestry Management for sustainability of the local growers. Yet, industry in the West is pushing hard for these supposed Clean Credits so they can avoid costly Carbon Controls in dying industries in the Western world.

Big cattle companies consent to halt Brazilian deforestation. Greenpeace won a major victory over Climate Change and deforestation in 2009 with the four world's largest beef producers consenting to ban the buying of cattle from recently clear cut land in the Amazon River Basin rain forest of Brazil.

Bertin, BJS-Friboi, Moffig and Minerva consented to Greenpeace to stop Brazilian deforestation.

The beef companies will control and observe their supply chains, requiring goals for registering farms supplying cattle. Also included is a ban on buying from farms that use slave labor, indigenous people and protected areas.

In June 2009 Greenpeace published a study "Slaughtering the Amazon". It prompted multinational corporations like Adidas, Nike and Timberland to threaten to cancel contracts that did not have requirements the products are not from cattle or slave labor in clear cut areas of the Amazon. Major beef buyers like McDonald's and Wal-Mart also required producers to guarantee similar provisions.

Brazil has the globe's biggest cattle herd. It is the biggest beef producer and exporter. Brazil is the 4th biggest Greenhouse Gas emitter in the world. The Tropical Rain Forest clear cutting accounts for nearly 20 percent of worldwide Greenhouse Gas emissions on an annual basis.

The Greenpeace victory was major.

Climate Change Control Action in Europe

The European Union changed its Carbon Trading program in 2010. The EU has the only Federal functioning international Carbon trading program in the world. The Kyoto Protocol called for treaty implementation once countries totaling 55 percent of the world's Carbon emissions ratified the treaty. Russia, a large Carbon emitter, joined the pact nearly 10 years ago after President George W. Bush made it clear the U.S. would not join. When Russia joined the treaty took effect.

In the Great Recession EU authorities see Carbon trading as a way to raise government revenue. Meanwhile Climate Change Control bills in both U.S. and Australia are facing difficult passage. A new government in Japan wants to cut Carbon emissions by 25 percent the most aggressive of any country in the world. Like the European Cap and Trade program, it faces stiff business opposition.

Carbon prices fell in Europe after it said it would not intervene as much as it has in the Carbon Market after 2012. The arrest of seven people outside London for failure to pay the value added tax on Carbon Trades caused a furor over fraud in the program and government intervention.

In a move to make the program more appealing to business the EU proposed to give a way a large amount of Carbon Allowances to industries most vulnerable to international competition while still achieving Carbon reductions.

France and Germany are proposing a "border adjustment tax" on imports from countries without mandatory Carbon reducing Cap and Trade programs.

The big fear in Europe is tariffs on imports from countries without binding Cap and Trade programs could start a major trade war. President Obama shares the sentiment. Ten Moderate Democrat Midwest coal producing state Senators blocked a bill in the Senate without tariffs on imports from countries lacking a mandatory Carbon reducing Cap and Trade program. The Senators feared loss of more companies oversees to unregulated countries.

The European Union announced a plan to invest in solar and carbon capture in 2009. The European Union plan aggressively cuts Greenhouse Gases. The EU is investing $50 billion Euros into research and development of solar and Carbon capture at coal plants.

The EU is comprised of 27 members. It already has a Carbon limiting Cap and Trade program but it is viewed as too expensive by business. Some EU countries have a Carbon Tax also.

Solar power is getting $23 billion Euros over the next decade in investment. Carbon capture is receiving $13 billion Euros over the same time. The hope is to have Carbon capture installed on all new coal power plants built after 2020. The Smart City plan aims at urban efficiency. $11 billion Euros goes to buildings and transportation to reduce their Carbon footprint.

$9 billion Euros will be invested in bio-Energy to make electricity and fuels from organic waste and plants. Nuclear power receives $7 billion Euros to develop nuclear fission, improve reactor safety, create less radioactive waste and mitigate the damage nuclear plants may cause.

Europe still relies on 40 percent of its power supply from coal burning power plants. The investment in Carbon capture is to catch up with the U.S. and Canada who are already investing billions of dollars in funding and other tax credits for Carbon capture projects.

Oil rich Russia's role in Climate Change. Russian President Dmitiri Medvedev and other Russian leaders gathered to discuss long term Energy policy in Russia in 2010.

He acknowledged that hydrocarbon rich Russia would need new, alternative sources of Energy. Russia is the largest exporter of natural gas and the second largest supplier of petroleum.

Russia is a large Energy user, mostly due to heavy industry, and old inefficient Soviet-era buildings in a cold climate. Russia's Carbon inefficiency is more than double the rest of the world and three times Japan's. Also, Russia's per capita Carbon emissions are increasing at a fast rate and will surpass the U.S. in 2030.

Russian participation in Climate Change Control is vital to the world economy. Based on the Kyoto Protocol, the baseline emissions inventory year was 1990.

Russia's economy caved in during 1992, along with its Carbon emissions. Russia had a huge surplus of emissions credits that it wanted to sell, billions of Euros worth, in order to make itself more Energy efficient. But due to a weak economy and abundant oil, efficiency gains have been lost.

If Russia uses more Renewable Energy, it can export more petroleum for a net economic gain. But new laws are required for Energy Efficiency, Clean and Renewable Energy, like wind and geothermal, both rich in potential in Russia.

Poor nations reject Carbon limits. President Obama brought a U.S. agreement to the Group of 8 Summit in Italy early 2009 only to be rebuffed by Europeans and rejected by poor countries. Europe wanted the baseline set to 1990, but the U.S. is using 2005 as a baseline date. The 1990 date would require steeper cuts sooner by the U.S. The Group originally had set a 3.6 degree reduction in temperatures by 2050, but ended up settling on a 2.0 degree decrease.

Meanwhile, China and India refuse to set Carbon limits even as their economies grow Greener. Less developed nations say the wealthy countries got to develop over 150 year period emitting Carbon free. The Developing countries feel they should be given the same chance to grow their economies unfettered.

Russia said that without China, India and Brazil on board the Carbon plan will go nowhere. The House Climate Change Control bill contains trade sanction provisions in it, but President Obama does not support them, saying they are protectionist and illegal.

But trade sanctions are the only way to get poor countries to participate in a Carbon Cap and Trade program.

Environmentalists in the U.S. were disillusioned with Obama's silence in 2009 during negotiation of the House bill. On the campaign trail many grass roots environmentalists worked hard for Obama and his tough talk on the environment.

Now many are planning retaliation against House Democrats who supported the bill. Many others felt Obama was going too far with Climate Change and healthcare during the recession.

The economy can not take such massive restructuring of some of its basic underlying pillars without prolonging the Recession many felt.

After many hiccups, the European Union (EU) Cap and Trade Carbon Trading program is at least starting to raise funds for the governments of Europe. Like the House Climate Change Control bill passed in June 2009, the EU gave away all of its Carbon Credits to industry for free when the program started.

Now the EU sells the credits. It is a tool the regional U.S. initiatives use with success. The Senate should also sell, and not give away, Carbon Credits. The EU expects to generate $26 billion U.S. per year through 2020 by selling Carbon Credits to industries.

Airlines were next in line to be brought into the Cap and Trade program in 2010. In 2009 there was still substantial fraud in the EU system, but less than in years past.

In 2009 the EU raised E $8 billion in U.S. dollars in Carbon sales. Germany made E230 million,

This will help offset the EU's contribution to the worldwide fund to help Emerging Nations cope with Climate Change impacts and implement new technologies to modernize their economies and control Carbon while developing.

The European Climate Change Control Cap and Trade Carbon Taxing system is almost a decade old. Now in Phase 3, the European model for Carbon Trading is replicating in Congress under Obama and the Democrats.

The first phase involves industry buy into a Cap and Trade program. Industry is given free Carbon Allowances of credits to pollute with Carbon. The Credits are given to the largest Carbon producers in the economy.

Then other sectors of the economy are included. After several years, the European's are selling the Carbon Credits for the first time, providing consumer utility bill rebates, Energy research funding and general funds.

As a national model, Congress and Obama are wise to follow the European model. Industry and Right of Center interest's support of the bill are crucial to the Cap and Trade program launch. The program can grow, adjust to economic cycles, expand to other sectors and provide a quantifiable measure of the external social cost of fossil fuel Carbon pollution.

Passage of the Congressional bill by the end of the session was uncertain in 2010. Mid-Term elections in November erased desire for another massive Federal government program.

Energy and Climate issues are very personal, emotional, controversial and political. It was unclear after an initial flurry how the silent 10 Moderate Midwestern coal state Senate Democrats would weigh in.

The Midwest has a voluntary state Carbon Control Cap and Trade system just launching. The posturing by Senator Jay Rockefeller (D-W.VA) is largely an election year, and reasonable regional constituent issue, measure of distaste for the inevitable.

The fact that Climate Change Control was not headline news gave me hope for passage. After major corporate member resignations in 2009 and 2010 over anti-Climate Change Control legislation espoused by the U.S. Chamber of Commerce, it became mostly silent and lobbied in the background. Immigration became the headline story in 2010, followed by the Gulf Oil Spill and the Iceland Volcano, all detracting from the legislation. Wall Street reform was not as prickly.

The bill, with the help of Lindsey Graham (R-S.C.), Olympia Snowe (R-ME), Susan Collins (R-ME) and perhaps another Independent or Moderate Republican, and Obama quelling the Moderate Democrat resource state Senators might have passed.

.

The U.S. has always pointed to Europe and the many things they do in Renewable Energy. Now many point to China.

It is not that the U.S. does not pay attention to Clean Energy. China is not a Free Market. The government directs investment in state run firms that are really corporations. Europe, more Socialist, is a Free Market economy. Yet Europe has high taxes and generous benefits. Even as the Euro slumped in 2010 it was still valued higher than the dollar.

The U.S. government and political system does not have the continuity of China or diversity of continuity as Europe. A policy switch in either direction by the U.S. government greatly affects the prospects of success of wind, solar, tidal and hydro. Bio-mass fuel is doomed by constant car technology improvements running on gasoline. Solar and wind, around since the 1970's Arab Oil Embargo for U.S. support of Israel, jump started Clean Energy in the U.S.

There is a life time of Congressional Energy Acts that never achieved anything in helping Clean Energy thrive and expand in the U.S. market place. In China and Europe the government strongly backs and mandates Clean Energy. Market barriers are reduced.

It is a policy question Europe, under different political stripes over the years, and China more recently, have consistently propped up with state investment and mandates.

Many in the U.S. clamor for such a U.S. action. Hopes were high for the new Energy Bill.

It was originally intended to simply establish a Carbon Cap and Trade program. The bill became a boondoggle of investment for the largest and most market self sustaining Energy sources in the U.S. economy. Clean Energy a vapid slogan of the Obama campaign was set back with no serious sustained investment.

Many on both sides of the aisle feel the government should not create and support new industries. This is despite promised jobs in many regions from labor to executives. Since it involves and affects every sector of the U.S. economy, huge sums of money, investment, jobs and lives are stake.

Change.

Sometimes too much too fast, can shock the system and body politic.

As much as I decry media hype, a member which I guess I am a part of, coverage is sparse, non analytical and mostly has a slant one way or the other.

I am not sure if in 2020 and 2030 we will see much of any more success of wind and solar, the grand daddies of Clean Energy since the 1970's, than we see now some 40 years after starting. It is easy to be cynical and dispassionate. The folks who work in all Energy sectors, new or old, deserve our respect and great admiration for giving us a great lifestyle. While complex our Energy intensive lifestyle provides us with a well functioning economy that is safe, protects the environment, provides us with security and allows us great personal mobility in our daily lives.

Clean Energy in the U.S. relies on your individual choice and action on which power to source your home and run your car. The consumer decisions you make about the labels you buy will influence corporate behavior far quicker and more responsively by management. This is true in the U.S economy and the world economy.

Market pressure will change carbon out patterns far more quickly and lastingly than a compromised legislative piece. Energy legislation almost always fails like its predecessors. We are left just as dependent on the fragile and volatile OPEC cartel and vulnerable to $5 gasoline for decades to come.

We are slow to change.

China, the largest Carbon creator in the world, leads the movement of the Emerging Nations on Carbon Control efficiency measures. Despite a Federalist state economy, China and the other top Carbon emitters India, Brazil, South Africa and Indonesia, are choosing voluntary Market Based Efficiency Measures. They reject the Western Command and Control regulatory structure, delegated by Legislators and Courts.

China is building more than 100 nuclear plants, practicing forestation, is a top world leader in solar panel building and exporting and has the largest wind turbine power supply in the world. All while it urbanizes, houses, feeds and employs peasants in the dozens of new planned mega-cities in the next decade.

India, more educated than China, is following a similar path to emergence from Colonialism. India is using voluntary measures, backed by urbanization, greater social services and spreading employment. Many medium technology devices are electrifying the rural villages with lighting, stoves, night time heating and television.

All the Emerging World Economy needs to do is offset Carbon creation from its urbanizing citizens as they enter the Digital Economy of the Western World.

It is the Western World, responsible for Global Warming, that is required to make lifestyle changes. The West must pay Carbon Taxes through efficient market based Cap and Trade Greenhouse Gas emission equivalent legislation and regulation.

The European Union and New Zealand have years of experience with this method. They are just now seeing job growth from the plan. The United States must follow suit.

The dominant Western Free Market was bailed out by a Marxist Economy. Yet the Corporate State is the world model of development, education and mobility. The West is no longer the leader, the only kid on the block. The Emerging World demands justice through Carbon Control Efficiency Measures legislated by Congress in the CLEAR Act by Senators Snow and Cantwell and regulated cost effectively by the United States Environmental Protection Act.

Chapter 7: The United Nations Framework on Climate Change

The UNFCC 2009 Meeting in Copenhagen, Denmark

There was pressure on U.S. to lead on Climate Change Control leading up to the December 2009 Copenhagen United Nations Framework on Climate Change (UNFCC) the successor meeting to replace the Kyoto Protocol. Returning from a G8 Summit, Obama was dealt his first major set back on the international stage. He left the meeting with no agreement or statement from the Group on Climate Change Control

This was a huge disappointment but not surprising. After nearly 20 years of inaction on the issue in the U.S. much had been expected. Europe rebuffed the U.S. for setting the baseline year as 2005 not 1990. This would have required steeper cuts in Carbon sooner by industrialized nations.

The 17 Developing Countries that were invited by Obama to join the Group of 8 in Climate Change Control talks were lead by India, China and Brazil. They rejected a 50 percent reduction in Carbon for them by 2050. The Western countries would have had to reduce Carbon by 80 percent.

There was time before the Copenhagen Protocol in December 2009 for the U.S. to take the lead on Climate Change Control world wide. A bill passed by the House in June 2009 would have reduced Carbon by 17 percent by 2020 and 80 percent by 2050 at a cost of $13 per household.

The Climate Change bill wound its way through the Senate over the summer. A key test of Obama's political skill and leadership was if he could hold the Northern Liberal Democrats and Southern Conservative Democrats together in the filibuster proof Senate. It was unclear in July 2009 how much more the bill will be weakened in the Senate. The Climate Change bill failed to pass. The Copenhagen Protocol, like the Kyoto Protocol, was doomed in the U.S.

The time was not right for the U.S. to take the leadership role on Climate Change Control world wide. The Europeans, while complaining, might have gone along with a U.S. plan if it was signed by Obama. Poorer nations might have been enticed to join once financing levels were set by the West to help Developing Countries make the transition to a reduced Carbon economy.

While Obama does not like it, the best way to force poorer nations to join a new Climate Change Control Protocol is through trade sanctions and penalties. This would greatly affect China, India, Brazil, Indonesia and South Africa.

In July of 2009, there was still a great deal to accomplish on Climate Change Control both at home and a broad.

Most Liberal environmental groups opposed all the compromises on the bill made in the House and were upset with Obama. Leading environmental groups like the Sierra Club and the National Resource Defense Council backed the bill.

They were seen as compromising with industry to at least get some tool in place to reduce Carbon soon. The world can not wait another 20 years to take unified action on Climate Change Control.

The challenge was there for Obama to whip the Democratic Senate to pass the bill. It was clear Obama could not wait on the sidelines. But he did.

President Obama had his attention on health care reform as did the major media outlets. Without leadership from the U.S., we were doomed to see another generation of increasing Carbon and rising temperatures.

Political barriers were rampant on the road to Climate Change Control. When the 1997 Kyoto Protocol covering 180 nations did not include China and India, there was political backlash in the U.S.

David Sandalaow was on President Clinton's negotiating team. He said going forward, "Only agree abroad to what you can implement at home."

Kyoto remains largely a failure. The U.S. never joined it. Members that did join have not met their Carbon reduction targets.

But if any agreement was to be reached in Copenhagen it would have to reflect the current American political climate.

Kenichi Kobayashi, of the Japanese foreign Ministry said the Kyoto framework must be abandoned for a totally new program, citing changes in the 12 years since Kyoto.

One of the biggest developments is that China has overtaken the U.S. as the world's top Carbon emitter. China and the U.S. contribute nearly half of global Carbon emissions. Copenhagen would not work without each of them coming to the table.

China, India and Brazil emit far more Carbon now than they did when Kyoto was agreed upon. The International Energy Agency says that 97 percent of increased Carbon will come from the Emerging World in 2030.

Such projections sway Moderate Senate Democrats like Evan Bayh (D-IN) to balk at legal limits on U.S. Carbon emissions. "I could not ask the American people to sacrifice and not solve the problem of global warming because the Developing World was not participating," he said.

India's Minister of Environment Jairam Ramesh said the U.S. short term Carbon reductions remain too low. "The stalemate in negotiations has not been caused by China and India," Ramesh said. "The make or break issue is emissions cuts. If there is no agreement on that, there's no agreement in Copenhagen."

The U.S. was slow to use provisions in the House Climate Change Control bill passed in June 2009 for a framework of the U.S. position at Copenhagen. These included such aspects as firm U.S. emission targets, investment and aid to Emerging Nations for forest preservation and Green Technology transfer.

Many top environmental leaders said because of the U.S. failure to act on Kyoto, aggressive action in Copenhagen was needed.

Carol Browner, Obama's Climate Change czar said it is "not likely" that the U.S. will have a Climate Change Control bill passed and signed by the President in time for Copenhagen. The U.S. will not endorse international goals the U.S. has yet to accept at home, according to Todd Stern of the State Department.

The alternative to break the stalemate going into Copenhagen under discussion by both the U.S. and India were nationally binding targets that would be capable of international scrutiny.

Jim Connaughton, a top Bush environmental advisor said that what comes out of Copenhagen may look a great deal like what Bush had promoted. "What countries came to realize after Kyoto was it was hugely problematic to have international environmental negotiations establishing domestic economic and Energy policy without first forging a domestic consensus," Connaughton said.

"What all major economies realized this time around is that they need to establish a domestic consensus on an agreed level of effort as a stronger basis for the commitments they make internationally and as a catalyst for international cooperation."

The promise of flexibility at Copenhagen signals not the end of discussion on Climate Change Control but the real beginning.

The Copenhagen Protocol negotiations noticed the absence of U.S. lead on Climate Change Control. When Obama was elected in 2008, Europe had great hope that he would lead the world to a solution on Climate Change Control.

In 2009, Europe was sorely disappointed in the new U.S. President. Mired in health care reform squabbles for months, Obama and Congress failed to take action on Climate Change Control.

Japan and Europe wanted a 20 percent or more cut in Carbon emissions from 1990 through 2020. The United States looked for a 2005 baseline foot print and a 17 percent reduction in Carbon.

China, India and Brazil remained noticeably recalcitrant in willing to make any legally binding cuts in Carbon. This has hamstrung U.S. policy and delayed preparations for the Copenhagen Protocol meetings.

Senator John Kerry claimed he would submit a bill to the Senate along with Barbara Boxer by the end of September 2009.

It was already too late.

A Carbon Energy bill was long overdue. The difficulty with health care in the U.S. Congress indicated it would be difficult to pass Carbon reductions.

My assumption was that Copenhagen would produce something lead by the Japanese and the European Union. What to do with China, India and Brazil and other Emerging Nations was be the biggest problem.

Considering nobody really expected anything concrete out of Copenhagen there was not much to talk about. The tiny island of Tuvalu, threatened by higher glacial melt seas, challenged industrial nations to make steep Carbon cuts and make deep financial commitments to poor nations. China and the U.S. waged a war of words, when really the two were not that far apart on many real issues.

President Obama and Secretary of State Clinton made great progress with India, China and Brazil in 2009. All three committed to going Carbon free as soon as possible. It required Western technical assistance, access to markets and financial support. The U.S. did not want to give China funding.

Some called Copenhagen a failure. Really it was a great mid-point for Obama's success in 2009. President Obama made real progress with three major emitters. The House and Senate did not neatly finalize legislation in time for Copenhagen. Such are the vagaries of domestic politics.

EPA has said it will regulate Greenhouse Gases (GHGs) with Operating Permits using the tried and true Best Available Control Technology (BACT) rules.

But there is no control technology for Carbon.

Ground injection is under investigation but years away from reality for any fuel, never mind coal, the test fuel of choice.

There was some clear progress at Copenhagen. Progress came out of Copenhagen the first week of the meting with a draft political document worked out not requiring binding Greenhouse Gas (GHG) limits but specific targets for the world's diverse economies.

The draft deal had the industrialized nations cutting Carbon by 25 percent to 45 percent over 1990 Carbon levels by 2020. The Developing World would cut Carbon by 15 percent to 30 percent over 1990 by 2020. By 2050 the world would see Carbon emissions cut by 50 percent to 95 percent over 1990.

The European Union pledged $3.6 billion to Developing Countries through 2012 to fight floods, drought and deforestation.

China verbally rebuked the U.S. for saying it would not finance China's Carbon reductions. Todd Stern, Chief U.S. State Department Climate Change negotiator said the U.S. did not believe China was an Emerging Economy because it has developed so quickly and is now the world's largest Carbon emitter.

The issue will probably be resolved by setting up private joint Carbon free ventures, Green Technology companies, and China-U.S. Climate Change Green Technology Research and Development Centers in China and the U.S. But China wants cash, and unlikely bet according to Todd Stern.

As the Copenhagen meetings progressed I came to realize why the Copenhagen Treaty was not needed. I will try to explain why I think a binding International Carbon Reduction Treaty out of Copenhagen in 2009 or Mexico City in 2010 or any other UN Treaty is not necessary and flatly will not work.

My focus came to me from my experience on lessons learned from the Clean Air Act of 1990, the Energy Policy Act of 1992 and the various transportation, highway and transit bills out of Congress since 1991.

I looked back at major EPA, DOE and DOT regulatory successes and failures for Criteria Pollutants like Ground Level Ozone, Carbon Monoxide, Sulfur Dioxide, Particulate Matter smaller than 2.5 microns, and Nitrogen Oxides. I applied the lessons learned to the Carbon Control debate.

The Clean Air Act was widely successful. The Energy Policy Act was largely a failure. The DOT funded both EPA and DOE transportation emissions and Energy reductions programs much to disgruntlement of the old highway builders. Domestic agency cooperation mirrors the China - U.S. split on funding Carbon reduction programs.

The love hate relationship between Big Industry and the Infrastructure Change Regulators, EPA/DOE/DOT is important to understand in looking at the future of Climate Change Control.

Industry always postures and fights verbally on TV against Clean Air, Green Energy and Technology and behavioral modification programs like telecommuting and car pooling. But then when the Northeast and California promulgate different standards for transportation sources of pollution, Energy use and targets for efficient appliances and utilities, the Industry complains to Congress that one National Federal Program is needed to reduce the cost of compliance with various regulations around the nation by different states.

The Federal Government and Congress can more easily set national targets for new technology pollution and Energy output and production than the states.

We have nearly 40 years of experience in Infrastructure Build Out, Clean Air and Oil Independence government experience. The government hardly ever solves public problems but is good at setting clear standards developed with industry, environmentalists and state governments to give industry the flexibility to solve the problem of air pollution, oil dependence, mobility, safety, and economic growth.

I have direct experience with EPA, DOE and DOT, Presidents Clinton and Bush and the Governors and Legislatures in Pennsylvania, Missouri, Maryland, California, New York, Massachusetts and Connecticut.

My insider's experience shows you how difficult and long it takes to write a government law, then pass Federal and then State government regulations. The interface of new Technology with Government is important, showing how industry solves public sector problems when the proper signal and non-Court room process works.

The free market will solve the problem of Climate Change.

Now we have Obama who is moving the Governments of the world in a clear direction toward Carbon reduction targets. They may always be voluntary and not binding.

EPA and Obama are working to foster Green Technology that must receive sustained private venture capital financing over the next century to succeed and successfully break down the numerous market barriers presented by the Carbon economy of the world.

In another clear victory for President Obama, India and China became official signatories of the voluntary Copenhagen Accord, joining 100 other countries in early 2010. The main term is to attempt to limit temperature rise by no more than 2 degrees Celsius or 3.6 degrees Fahrenheit over pre-industrial levels.

The West will donate over $100 billion annually to Developing Countries to help them adapt to Climate Change, bringing Low Carbon and Advanced Technologies to them.

China and India follow a "Carbon intensity" measure reducing Carbon based on per unit economic growth. China proposed to curb intensity by 50 percent over 2005 by 2020. India proposed a 25 percent cut, excluding its vital agricultural economy.

The U.S. proposed a 17 percent cut in Carbon by 2020 over 2005.

Will international, voluntary Climate Change Control measures work? The history of voluntary government and corporate effective use of voluntary controls on almost everything, have never worked, well or long.

The best example is the Blue Skies program of President Bush that allowed voluntary monitoring and reporting of air pollution by corporations with little EPA oversight or enforcement of law. It did not work.

Now that Obama is enforcing air pollution laws and issuing fines. This is forcing environmental protection back into corporate budgeting. It is causing job loss in Texas, the petro-chemical and heavy industry states.

Will China, India, Brazil, South Africa, Indonesia and other leading Emerging Economies pledge to use voluntary Climate Change Control measures work?

No.

But the free market might.

The US, to the disappointment of Europe and New Zealand, will also end up with voluntary Carbon Control programs. That is except in the regional Greenhouse Gas control states, many, in the Northeast, West, Midwest and Canada.

China is following through on heavy investment in nuclear, wind and solar. India faces a future of Clean Development to ease poverty, modernize and electrify rural regions. Brazil is working more on forest protection, as is Indonesia. The status of South Africa is unclear.

Government planning and funding is important in stimulating new, Clean Technology. But the private venture capital investors are the ones who ensure long term market barriers to Clean Technology and Development in the next several generations are overcome.

The UNFCC 2010 Meeting in Cancun, Mexico

The United Nations Framework on Climate Change (UNFCC) met in Cancun, Mexico during the end of November through early December 2010 to hammer out further progress on the Kyoto Protocol passed in 1997. Many of my colleagues that I have spoken with in Europe and around the world since the failed Copenhagen talks in December 2009 in Denmark agree that a sector by sector approach to Carbon Control may work best in the near term to reign in the world's increasing Carbon production.

This could take one of many forms as the nations of the UNFCC gathers at meetings in the coming years. The UNFCC could follow the United States lead and pass a measure requiring that any source that emits 25 tons per year or more of Carbon Dioxide equivalent (CO2e) gases be controlled and reduce Carbon output. The EPA is doing this under a Court Order imposed by the United States Supreme Court in 2007.

This is going to happen regardless of whether the Republicans take the Presidency and full Congress in 2012. Many legal scholars believe the Roberts Supreme Court is the most Conservative Supreme Court in the recent history of the United States. The Roberts Court continues to stand by its 2007 decision saying that Greenhouse Gases or CO2e are indeed Criteria Pollutants covered by the Clean Air Act of 1970, 1977 and 1990.

Alternatively the UNFCC could follow China's lead and require that a certain number or percentage of the world economy is covered by Carbon Control regulation. Compliance options would be presented to an independent auditor like Infosys of India or Deloitte of the United States. The auditor would verify that the reductions are real, verifiable and surplus.

I suggest the UNFCC require the top 30 percent of the world's Carbon sources have commitments to reduce Carbon by 20 percent by 2020 and 80 percent by 2050. This would capture the top utility, cement, glass, steel, automobile and transport producers around the world. This will cut world Carbon production by 50 percent to 75 percent in ensuing years.

The UNFCC could phase in sectors of the economy covered by the rules but they must be harmonized so that one nation's export economy is not placed at unfair advantage against on non-cooperating country. In this rare instance, trade sanctions are warranted against non-compliant countries who seek to exploit the nature of Treaty agreement adoption around the world. Europe is already levying tariffs on imports from non-Kyoto Protocol countries.

Such creative thinking and leaving verification to hired contractor may be the best way to solve the slow progress on Climate Change Control worldwide. Every industrial and rapidly Emerging Country in the world sees Clean and Renewable Energy as its key to the next century in terms of economic growth, development, mobility, national security, independence and a cleaner environment while reducing Carbon.

Clearly the time has come for the world's nations to acknowledge that Clean Renewable Energy is here. The world's nation must begin passing sector by sector controls of Carbon until the world has a less Carbon intense economy with more Renewable Energy.

The United Nations Framework on Climate Change (UNFCC) wrapped up two weeks of international Climate Control talks in Cancun, Mexico in December 2010. There were no expectations in 2010 after the failure in December 2009 in Copenhagen, Denmark. No agreement on Carbon reductions was even discussed. The major media described this year's talks as focusing on secondary issues.

The only agreement signed was a re-affirmation of the pledges by Developed Countries to help Developing Countries between now and 2020 with $100 billion in funding for Green Technology. It is unclear whether this means Renewable Energy or Clean Energy. Some fossil fuels do produce less Carbon than others.

There were conflicting reports in the press about talks between China and the United States. The two are the number one and two world's greatest Carbon polluters. The reports were about verification of Carbon reductions. There were a few different terms discussed such as an International Consultative Analysis approach whereby a third parry would verify the Carbon reduction of each nation.

Other terms were mentioned but essentially it means quantification, verification and non-duplication.

It appears China who had resisted for several years any outside verification of Carbon cuts is now accepting non-governmental verification of its Carbon cuts.

The bilateral talk between China and the United States was the biggest story out of 2010 Climate Control talks. China is now at least opening up to some form of outside verification of Carbon Control programs. This is probably a public relations move by the Chinese government. They vociferously denounced international meddling in its detention of the 2010 Nobel Peace Prize winner. While China still has a strong distaste for international scrutiny of it s Human Rights and child labor politics, it appears to have softened on verification of Carbon cuts.

China is making very real commitments to Carbon reduction programs. It closed hundreds of old and inefficient coal and manufacturing plants around the country in 2009 and 2010.

In the United States, various coal power generating plants closed replaced by either natural gas powers plants or more efficient new coal power plants.

Diesel power plants are rapidly disappearing as the price of a barrel of petroleum could hit $150 by late 2011. Many analysts have long predicted $5 per gallon gasoline as the new norm in the United States.

Multilateral Climate Control talks have appeared to make no progress. It is too complex. The voice of a lone dissenter lost in another century only can spoil the process.

The Kyoto Protocol countries do not want to commit to more Carbon cuts without the United States, China, India, Brazil, Indonesia and South Africa committing to binding Carbon cuts. The Kyoto Protocol countries only make up 20 percent of the world Carbon pollution.

So while the UNFCC will continue to meet over the next several years they are really a toothless and purposeless body. Real action must happen between the United States and China, Brazil and India and Indonesia and South Africa.

Progress has been made in forestry management worldwide, Renewable and low Carbon Energy, but much more needs to be done. Fortunately the real power is one on one between countries and bilateral agreements.

The multilateral process has failed in a world mired in job loss and austerity. The developed nations can barely feed and employ their own residents, never mind donate $100 billion to poor nations.

The UNFCC Requires Improved Fuel Economy Standards

The future role of transportation could be an important key to future UNFCC negotiations in years to come. The United Nations Framework on Climate Change (UNFCC) should hammer out the next steps in Climate Change Control for the world.

The United States began phasing in a light duty passenger car Corporate Average Fuel Economy (CAFÉ) of 35 miles per gallon in 2011 up from the current fleet wide average of 25 miles per gallon. Heavy duty vehicles over 12,500 pounds gross vehicle weight are also controlled.

The current gallon of gasoline emits about 400 grams per mile of Carbon Dioxide plus Methane (CH4) and Nitrogen Dioxide (N2O). The 2015 level will be around 200 grams per mile of Carbon Dioxide.

The UNFCC should adopt light duty passenger vehicle fleet wide CAFÉ requirements of 50 miles per gallon for 2030 and 100 grams per mile of Carbon Dioxide. The goal for 2050 should be zero grams per mile of Carbon Dioxide for light duty passenger vehicles. By 2050 hydrogen and electric vehicles under 12,500 pounds gross vehicle weight should be readily available at competitive market purchase price and the electric and hydrogen charging station issue will have plenty of time to get resolved.

The story for heavy-duty trucks over 12,500 pounds gross vehicle weight is different. It will take much longer to introduce electric battery and hydrogen fuel cells that efficiently provide enough power to freight transport for buses and commercial trucks. For this reason, a different measure needs to be applied. The UNFCC should double the current heavy-duty truck CAFÉ average from approximately 10 miles per gallon to 20 miles per gallon by 2020. The goal for 2035 should 30 miles per gallon gasoline equivalent. The goal for 2050 should be set at 50 miles per gallon.

Since there are many other fuels in the heavy duty truck and bus transport sector the miles per gallon fuel economy should be translated to total Carbon Dioxide Equivalent (CO2e) emissions. This would mean that diesel, natural gas, propane, bio-mass fuels mixed with diesel and hybrid electric diesel and other fuel engines would have to achieve the same CO2e emissions output for the gasoline equivalent miles per gallon instead of only the same miles per gallon.

For example diesel may be able to meet a gasoline gallon equivalent of 20 miles per gallon by 2020 with very little manufacture engineering changes. But if the length of time the CO2e emissions remain in the atmosphere is translated into the to the 20 miles per gallon gasoline equivalent, natural gas with Methane (CH4) as its primary component will achieve the 20 mpg CO2e emissions threshold far more easily than diesel or any of its blends and hybrid variations. Indeed the heavy duty engine of the post 2050 era may well be a natural gas hybrid electric engine.

What remains clear is that gasoline and diesel are toxic both in terms of Criteria Pollutants that cause Smog, Acid Rain, Carbon Monoxide poisoning and Climate Change. Natural gas is a promising bridge to a Carbon Zero transport future. The goal may be hydrogen and electric heavy duty trucks in the next century, but until the technology is there, natural gas heavy duty engines are proven, safe, efficient, market ready with expanding fueling stations and far less deleterious human health and property than gasoline or diesel.

The UNFCC should consider debate and adopt these reasonable standards.

Chapter 8: Renewable versus Clean Energy

Clean Energy and Renewable Energy are two terms that are used interchangeably often meaning the same thing. First let me say that to qualify as a Renewable Energy source, it must be clean on both Criteria and Greenhouse Gas emissions. But Clean Energy and Clean Fuel is not always, sometimes hardly ever, Renewable. I speak here of Clean Energy as the technology enhanced affects of fossil based fuels, most notably natural gas, but also nuclear, petroleum and coal.

First of all, I may have just lost most readers who are most concerned with Renewable Energy. Bear with me here as I try to explain what may seem like flawed logic.

Conventional Energy like coal, petroleum, nuclear and natural gas are the current Energy market share leaders. These fuels are not going to sit on the sidelines and watch their profits and industries disappear over the arrival of Renewable Energy.

Indeed, coal, petroleum, nuclear and natural gas firms should be the very firms supporting wind, solar, tidal, hydro, hydrogen and biomass Energy as the future of the World Economy. Conventional Energy companies have the resources to invest in, research and develop market ready Renewable Energy facilities around the world.

It is a win for the Traditional Energy firm's public relations wise and politically. It is a major win for the struggling start up Renewable Energy firms around the world.

This may seem to make no sense to the Alternative, Renewable Energy enthusiast who sees all fossil fuel as essentially evil and with no role in the New Renewable Energy World Economy. This is short sighted and serves no interest but a base emotional desire to believe that all Renewable Energy is pure.

Renewable Energy has some of the same social externalities of pollution, disease, and property loss associated with Traditional Energy. It is unwise to believe there not already enormous problems siting wind and solar farms. It is unwise to think that new and other social harms from Renewable Energy will not be brought to light in fully vetted and well rounded Environmental Impact Assessments.

Natural gas is the winner as he bridge fuel to the Carbon zero economy of the future most people seem to agree is the goal. Natural gas in most of its field wells has zero Carbon Dioxide (CO_2). Natural gas' main pollutant is Methane (CH_4) which only remains in the atmosphere for several months as opposed to Carbon Dioxide (CO_2) which lasts several hundred years in the atmosphere. Great Britain achieved its Kyoto Protocol Carbon Reduction Targets by simply switching the national economy off of coal and onto natural gas.

Fuel switching from coal and petroleum to natural gas is not easy in the world's two biggest Carbon polluting countries China and the United States.

China has ambitious plans to build dozens of large coal and nuclear power plants to fuel its increasingly urban population. Once rural Chinese residents are living in cities and suburbs now and many more soon will. They are demanding the same standard of living and quality of life as Americans and Europeans.

The shear sums of invested capital, money and investment involved in coal, petroleum and nuclear make them likely players in the World Economy for decades to come and most likely well into the next century.

Pure economics will dictate that indeed when measured across all social externalities, pollution, disease and property damage costs that Renewable Energy is cheaper to produce and use than Traditional Energy. The Traditional Fuels will eventually be priced out of the market place.

This will be come more obvious in coming years as more and more nations around the world start pricing the societal cost of Carbon into Energy use either by legislation or regulation.

This is happening now in China and the United States through Court Orders and Executive Orders. Legislative mandates may never be needed accept to eventually pass nationally world wide binding treaties on Carbon reduction that if not adopted place the host country at a t severe economic disadvantage.

Renewable Energy is on the way. Traditional Energy will develop and deliver it. Natural gas will be the last market viable Traditional Fuel in years far ahead.

Should Fossil Fuel Be Banned?

The whole oil drilling catastrophe off of the Texas coast raises the question of the need and continued use of fossil fuel in any form. 11 men dead, thousands of endangered fish and birds killed, fishermen economically dislocated, tourism impeded, swimming water quality unhealthy.

All for petroleum we can not afford to fill our tanks with because the price is controlled by the New York Mercantile Exchange (NYMEX.com) speculators and not the Capitalist law of supply and demand.

NYMEX changed their website domain the summer of 2010 to the controlling private for profit speculators innocent sounding name CME Group.com. The web page is CMEGroup.com but do not let them confuse you. This is the bastion of Capitalist corruption and speculation.

These Ponzi boys are far worse than Inner City drug dealers. You need petroleum. BP, ExxonMobil, Shell, Chevron, and the smaller petroleum players are planning on $6 per gallon gasoline by the summer of 2012 to drive President Obama and the Democrats out of power.

Many analysts say gasoline at the pump will reach $5.25 per gallon in 2011, causing a severe Worldwide Great Depression. The petroleum industry is scheming to recover oil industry clean up costs of the Gulf of Mexico Deepwater Drilling Platform Oil Spill Catastrophe of 2010.

The argument to go cold turkey off Carbon rich fossil fuel is clear. The costs, even without Global Warming counted, are not worth the benefits. You can make the argument that our troops are dying for $4 per gallon oil.

Going fossil Energy free is quite achievable, requiring efforts similar to the Manhattan Project and the quest for the moon. More importantly, a massive government Clean Deal creates hundreds of thousands of New Energy Technology careers in every district, from R&D to sales, installation, operation and maintenance.

Why not? We have tried everything else and nothing works. Our fossil fuel addiction is killing us and our troops, sacrificing our economy, environment, safety, mobility and security. Outlaw fossil fuel today. Pass the Clean Deal.

Building a Bridge to a Clean Economy

What can we do today and in the ensuing years to build a bridge to a cleaner world economy? This is a simple question with no easy answers. My postulate is to follow Great Britain in the near term. Great Britain changed its power generation for households and industry from a core coal base to natural gas. This simple yet complex switch enabled Great Britain to meet its Kyoto Climate Change Control commitments ahead of schedule with a net job gain.

Such a move is not so easy in the United States and China. What Great Britain tells us is that the way to a lower and Carbon free world economy is through Carbon fossil fuels themselves, albeit lower generating intensity Carbon fossil fuels like natural gas and propane. In other words we need to incrementally step down to lower Carbon output fuels that are indeed fossil fuel based. In doing this we will reduce Carbon output with the least cost compliance option to the world economy.

Natural gas and propane are proven fuels in many transportation and utility generating power plant engines.

The goal is to get to a Carbon Free economy without damaging the economy. This means coal, nuclear and imported petroleum will remain vital to world economy even as we try to displace it from its preponderant use. This means that fossil fuel companies must invest in Clean and Renewable Energy projects like wind solar, tidal and hydro and the transportation fuels of hydrogen and fuel cell battery technology. This makes a great deal of sense. Fossil fuel firms must invest in the answer, the successor technology that will someday supplant our finite coal and oil reserves, however near or far this may be in the future.

This is the goal of the next several decades. With or without Federal legislation in the United States and China fossil fuel companies must be the ones to diversify Energy sources and slowly wean the world off of Carbon dependence. This will not be easy but we must buck up and accomplish the formidable task.

Climate Change Professionals Need Classes in the Clean Air Act

In transportation Energy is called Fuel. As someone who has worked in Energy Conservation early in my career and then in transportation, air quality and alternative fuels since 1990, I can speak to Climate Change Control professionals. Climate Change Control professionals must learn transportation and the history of the Clean Air Act of 1970, 1977 and 1990. If you are working in Clean and Renewable Energy for Climate Change Control you must know what CAA means.

I urge all Climate Change Control professionals to take a class in the history of the Clean Air Act. You should know why Senator Ted Kennedy enacted PSD and what it means (Prevention of Significant Deterioration), what the National Ambient Air Quality Standards are, and the various divisions of the EPA, DOT and DOE as well as their State counterparts.

You should know what Attainment, Maintenance and Rate of Progress means. You should know what mode, queue, signal timing and fleet means if you are a utility person. If you are a transportation person you should know what a joule, erg, volt, kilowatt and megawatt are and what a BTU is. All should know the difference between a short ton and a metric ton. Yes, air quality planners round neatly to 2,000 pounds, no European infiltration allowed here.

Chapter 9: Climate Change Control Strategies

Coal Control Strategies

A Carbon capture pilot program in West Virginia tests and evaluates the new technology. In New Haven, West Virginia a crucial experiment is underway that may determine the fate of the coal industry worldwide. The technology tested is Carbon sequestration or Carbon capture. The technology takes Carbon Dioxide emissions from the coal plant, mixes it with a chilled ammonia compound, and buries it thousands of feet below sandstone in a dolomite formation.

There are numerous problems with the untested and unproven technology. The Carbon capture and mixing factory consumes between 15 percent and 30 percent of the Mountaineer coal plant's Energy output, making solar and nuclear more cost effective.

The cost of the Carbon capture factory is half that of the total cost to build Mountaineer. The project cost over $100 million with American Electric Power, the plant's owner, paying $73 million. During the pilot phase only 1.5 percent of the plant's Carbon is buried. If it proves to be cost effective the utility could bury 90 percent of the plants Carbon. The jury is still out.

Environmentalists say Carbon injection will cause ground water pollution and will leak.

Much is riding on the outcome of the Mountaineer pilot project. This is true for both supporters and detractors of clean coal technology.

President Obama restarted a clean coal project in 2009. After President Bush ended the project in 2008, the Department of Energy is restarting a clean coal project in Illinois. The project called FutureGen, liquefies and gasifies the coal, sequestering the Carbon Dioxide and injecting it deep underground.

The U.S. gets half of its electrical supply from coal. The project will cost $1.5 billion with $1.1 billion coming from DOE and another $400 to $600 million coming from FutureGen, an alliance of coal industry utilities and groups.

The project is seen as an important way to tap the country's vast coal reserves. But environmentalists claim that coal is too dirty no matter what method is used to try to clean it.

We already store spent nuclear fuel and natural gas under ground. The technology could be exported to China creating thousands of jobs in the U.S. and China.

China's economy is expected to grow exponentially through 2020 as the rest of the world grows slower. China was lagging on negotiations in 2009 leading up to the Copenhagen Protocol. It called for highly unrealistic Carbon reductions of 40 percent over 1990 by 2020 by the U.S. China has invested heavily in Clean Technology over the past few years. China has conceded a low Carbon economy is the future for the world.

In the spring of 2009 House Republicans introduced an Energy bill with outdated and ineffective ideas. The plan calls for 20 new nuclear power plants by 2020, increased oil and natural gas exploration on public and private lands and offshore water areas.

The increased revenue from the new leases would be used to fund Clean Energy technologies like solar and wind. The Republicans oddly enough do not include clean coal technology.

These are the same ideas Republicans have been mentioning for the past decade with no implementation during their time in power. It appears the Republicans are just posturing with the public to present a plan counter to Waxman/Markey.

The Republicans called the House Climate Change Control bill national Energy tax at a time of great economic turmoil. The House fast tracked the Waxman/Markey bill calling for a Carbon Cap and Trade national program. It would have given away 85 percent of the emissions allowance for Carbon.

Speaker Pelosi fast tracked the bill to the floor of the House when she realized health care reform would take up a great deal of time. There was very little media coverage of the Carbon Cap and Trade program and not much public debate.

The costs of a Cap and Trade bill start at around $3 per month per household in the early years. The cost eventually climbs to $10 to $15 per month for Energy bills in outer years.

The bill aims for a 17 percent reduction of Carbon by 2020 over 2005 and 85 percent reduction by 2050. In contrast, the Republican Bill offers no mandatory Carbon reductions at all.

Coal is likely to remain a large part of the Energy future. Midwestern coal state Senators of both parties have ensured in the Climate Change Energy bill that coal will remain 50 percent of the U.S. power supply well into the future, with the addition of Carbon sequestration.

This notably displaces natural gas as a near term quick Greenhouse Gas reducing fuel source. Members of natural gas lobbying groups are marshalling resources to ensure natural gas will grow as a fuel in the power system, but coal is a clear winner.

The only problem is Carbon injection into the ground, or sequestration, is largely unproven. A Portsmouth, NH company, PowerSpan, has a project running in Ohio, but there are few others. The Energy bill does a great deal to aid Carbon injection research, development and deployment,

The bill cited China as a main market to export the technology. The horse trading continued in Washington, DC, watering the Waxman Markey House bill down more and more in the Senate suiting it to fit existing fuel source constituencies around the country. By November 2010 prospect dimmed in DC that a Carbon trade bill would pass.

The health care reform rancor obscured Climate Change Control from public focus and the press.

Old coal plants are still a problem in much of the United States. Coal plants pollute the air, public water drinking supplies and are rarely enforced against by State and Federal air quality regulators.

Besides causing Climate Change old coal fired plants pose a public health risk to immediate local environs. They are extremely uneconomical and inefficient. The 85 year old Crawford Generating Station in Little Village Chicago, a Mexican community, is an example. The Fisk Generating Station 6 miles away, built in 1903 is another example. These plants produce high, local concentration of PM2.5, a very harmful lung carcinogen.

Many public health officials say Fisk and Crawford are poster children of Clean Air and Climate Change need for reform. These legacy plants were exempted by the Clean Air Act of 1977 from being required to use the Best Available Control Technology (BACT).

Brian Urbaszewski Director of the Environmental Health Programs at the Respiratory Health Association of Metropolitan Chicago said, "What we're dealing with here is the Cuban auto fleet - a bunch of facilities built in the 1950's and 1960's that are continuing to be rebuilt over and over that's not the way the law was intended to work."

Proponents of closing Crawford and Fisk hope that the current Climate Change bill before Congress will force the two plants to close. The Climate Change bill could raise the cost of building new power plants. Utilities still try to squeeze out more production from inefficient, old, dirty power plants.

The Waxman / Markey bill puts no limits on existing Carbon producing plants.

Power plants produce Ground Level Ozone and PM2.5 both powerful lung irritants, especially for the elderly, young, outdoor workers, and people with asthma and cardio-pulmonary heart lung diseases. Fisk and Crawford are a public health hazard. If Waxman / Markey or some version of it passes the House and Senate, Fisk and Crawford most likely will close, long over due.

New coal power plants are under construction across the Midwest and South despite the economics that make coal a loser both for the investor, public health and property. According to Matthew Wald of the New York Times coal power plants are springing up everywhere in the Midwest and South with barely a scant account of a Carbon tax and the human cancers caused by coal emissions.

The Prairie State Energy Campus outside of the St. Louis suburbs in Illinois is the site of a $4 billion coal plant. While both the Northeast and the West Coast pursue cheaper natural gas fired Energy plants and investment in Renewable Energy like hydro-power, wind turbine power and solar panel power, the Midwest and South are going with an economic and public health killer.

The Prairie State coal plant has a 700 foot smokestack. This is illegal under EPA regulation and laws passed by Congress banning enormously high smokestacks. Very tall smokestacks carry the soot and pollution hundreds of miles away to the populated East Coast.

Apparently the state of Illinois neglected this little fact in the plants Title IV Operating Permit. The plant is illegal and should be shut down. Why the tall smokestack? The coal industry knows local towns would not tolerate the coal dust and ash from a smokestack that met the Federal law. So the Prairie site broke the law and should be fined out of existence.

Mr. Wald points out that Prairie State will be belching out Carbon well past the 2050 deadline to cut Carbon emissions nationally by 80 percent. Plant managers doubt that a Carbon tax will have any effect on them. Coal power can sell as cheaply as 5 cents a kilowatt hour. Even a Carbon Trading price of $25 per ton would only add a 2.5 cent per kilowatt hour cost to the price of coal.

Elsewhere in the country cheap natural gas prices and its plentiful supply have severely slowed the construction of all other types of power plants, especially coal and nuclear power plants.

Duke Energy is building several coal fired powered plants that will, like Prairie State, use coal gasification. Coal gasification is the crushing of coal and then combusting it, making it easier to control emissions. Some of Duke's plants in North Carolina and Indiana will then pipe the Carbon pollution to oil fields in Texas and the Gulf for underground burial and injection into to oil fields to help improve oil yields.

This all sounds well and good, but Duke only plans on sequestration or underground burial if the price of coal power gets too high because of a Carbon tax.

This is unlikely. It is also a new an unproven technology with very high costs that make it very unlikely to ever be used, except in a clean coal TV advertisement.

While new coal power plants are under construction in the South and Midwest, many large electric power companies like Exelon are switching to natural gas. Natural gas prices have plummeted in the past two years as huge amounts of domestic shale are now having natural gas extracted from them. There is reported to be hundreds of years of domestic natural gas available.

While "fracking" or fracturing of the under ground shale natural gas deposits is controversial, it continues in widespread use. Many residents report ground water contamination and their wells poisoned from fracking. Hopefully the practice will be refined or at least moved away from sensitive local populations who rely on well water

With the price of natural gas low and stable, electric power generating companies see natural gas as a cost cutting and effective way to meet numerous environmental regulations from the EPA. New Sulfur Dioxide, Nitrogen Oxides and Particulate Matter regulations from U.S. EPA require extensive and costly new pollution control on coal power plants.

A majority of the coal power plants in United States are older than 50 years and many are almost 80 years old.

Like China, United States electric power generators are choosing to shut down old, inefficient coal power plants and replace them with new coal, natural gas or nuclear power plants.

China has chosen a path of new coal and nuclear power plant construction while a majority of the United States electric power generators are closing old coal power plants and building combined cycle natural gas power plants.

Natural gas is innately and inherently the cleanest fossil fuel known to mankind. It has virtually zero Particulate Matter or soot traditionally associated with coal power plants. Several inefficient old coal power plants can be replaced by one new super efficient and ultra clean natural gas power plant. One electric power generator saved a half a billion dollars and many others save hundreds of millions in fuel switching from coal to natural gas.

Natural gas also has almost zero Sulfur Dioxide in it. Its Nitrogen Oxide emissions are extremely low and controllable with low NOx burners. This controls the combustion temperature by spreading out the flame. It is sprayed with a water mist for peak Energy production and minimal nitrogen oxide emissions.

Volatile Organic Hydrocarbons or VOC, primarily Methane in natural gas are also very low and easy to control with existing low cost emission control technology. In addition, Methane has a half life in the atmosphere of months compared to the hundreds of years for Carbon Dioxide and Hydroflurocarbons (HFC). Carbon Monoxide is not a major concern either for natural gas.

The electric power generators see the combined economic impact of cheap and plentiful domestic natural gas. It sees the very low air pollution emissions for traditional Criteria Pollutants covered by the Clean Air Act and Greenhouse Gases which the Roberts Supreme Court ruled is governed by the Clean Air Act also.

Averaging, Banking and Trading, or popularly called Cap and Trade was created by economists in the Carter Administration and embraced by the Reagan Administration. Bubbling and Banking emissions credits from new power plants can be sold to other dirtier power plants or banked by the owner for future increased production and plant expansion.

The same concept of bubbling and banking is used in the fleet wide Corporate Average Fuel Economy standards that allow SUV low mileage and high emissions to be offset by a higher percentage clean and ultra clean, high mileage passenger cars. Now, we simply need to realize that natural gas for transportation is the next step for the economy.

Is there a role coal in a Renewable Energy Economy? Coal contains dozens of known carcinogens to animals and humans. Coal air pollution causes Acid Rain, Ground Level Ozone (O_3) pollution, emits large amounts of both Carbon Monoxide (CO) (a local asphyxiate) and Carbon Dioxide (CO_2). Coal unleashes deadly very fine Particulate Matter or soot smaller than 2.5 micrometers or the thinnest of human hairs. PM2.5 adhere deadly air toxics also found in coal like Toluene, Benzene, Xylene, Mercury, Arsenic Cadmium and dozens more.

The deadly carcinogenic toxins attach to the very fine soot and deeply penetrate the human lung. This is the source of lung cancer appearing in non-smokers under the age of 35 at an alarming rate, hundreds of miles from the nearest coal plant.

Yes, coal is dirty, cancerous and it also contaminates local ground water and public drinking water supplies. The most common form of coal extraction was dynamiting the tops of mountains in Appalachian coal country in a process called Mountaintop coal removal. The process was so harmful to both the natural and human environment, especially vanishing mountain area streams that it was banned by President Obama as quickly as he possibly could.

Yes, coal is deadly.

And coal is dirt cheap. Too cheap for the real cost not accounted for in utility bills for the natural and human life it kills and the property damage it causes. Coal should be illegal to use.

Yet coal is more likely to play a larger more enduring role than nuclear power in the world electricity Energy production supply well into the next century.

Coal will remain cheap even with a Carbon Tax, a human carcinogen and property loss assessment tax placed on it by the Federal government. Coal is simply the cheapest and most plentiful Energy source in nearly every major economy of the world.

It is easy to successfully and profitably continue to use Coal as a major Energy source for the world economy for several decades to come.

The Environmentalists, ever hopeful of killing coal use, recently released a major report about underground injection or so called sequestration of Carbon Dioxide (CO_2) into underground rock formations. The Environmentalists assert that the stored pollution will leak into ground water poisoning public drinking water supplies.

Regardless of the claim by the Environmentalists, early reports indicate that ground injection sequestration is simply too cost prohibitive to ever work on a large scale basis. As the technology praised by some as the answer that will allow coal to be a clean Carbon player in the future there are only handful of small pilot projects of underground injection sequestration around the world.

All of them indicate that it costs nearly as much to build the Carbon Dioxide (CO_2) gas capture equipment facility and inject it underground as the cost to construct and then operate the original coal fired power plant. If this technology were indeed more promising than industry public relations the coal industry would have been testing this technology for decades by now.

The coal industry is the most caught in the last century of any of the fossil fuels. It has taken the coal industry nearly 40 years to comply with some basic requirements of the Clean Air Act of 1970.

There are still hundreds of coal power plants especially in the Midwest and Appalachia that do comply with the Clean Air Act requirements nearly four decades after enacted by Congress and signed into law by President Nixon.

Coal is teflon. It is political gold, cheap, plentiful and often found and burned in remote areas of the world where the educated urban populace does not see it. Coal is dirty and deadly.

Coal unfortunately has enough political and economic muscle around the world to continue to break basic air quality law, never mind try to control Greenhouse Gases like Carbon Dioxide.

Coal is evil. Coal needs banning immediately. But unfortunately, coal will most highly likely continue polluting local ground water and cause air borne cancers in urban residents hundreds of miles from where it is extracted and burned for many generations into the future, even with a Carbon Tax, a Human Lives Lost Tax and Property Damage Tax placed on the deadly, noxious fuel.

Petroleum Control Strategies

Will we ever get off of our petroleum addiction? This is most highly unlikely. U.S. dependence on imported petroleum runs the world economy. Despite dramatic attempts by Obama and Congress, U.S. dependence on imported petroleum is likely to continue for the next hundred years.

The U.S. has already been trying for 40 years to wean itself off of petroleum with little success and few results. President George H.W. Bush and President Bill Clinton tried in the early 1990's to get the U.S off of petroleum with a combined Clean Air Act, Energy Policy Act and Intermodal Transportation Efficiency bill. The Clean Air Act was the only success. Energy efforts even modestly aimed at fleets failed to be implemented and enforced.

The long history of failure of the U.S. government to get the U.S. off of imported petroleum is more a market dictate than policy failure. Quite simply, without powerful government intervention, the market favoring centralized power generation and distribution can not be tilted by local solar and wind options.

There is more hope in the developing world where centralized distribution is simply not economical. It is believed that decentralized wind and solar in the Developing World could greatly reduce Carbon emissions.

But China and the U.S. are far apart on Carbon reductions with China calling for a 40 percent reduction in U.S. Carbon emission by 2020 over 1990. It is rumored the U.S. will only cut Carbon emissions by 4 percent between 2005 and 2020.

Several new studies in 2009 predict sea levels will rise two feet higher along the Atlantic Coast than the rest of the world by 2100. Perhaps we can rescue large urban areas but many seaside resort towns will be lost to the ocean.

While you may still be questioning Climate Change in your head, there is no disputing that the ocean has risen 0.7 inches per year for the past century.

In the end, the market will dictate the answer. When the human and property loss damage of Climate Change are felt on the world economy, the markets will react. But that is likely 100 years away.

While President Obama's promises are a sign of change, it is unlikely the U.S. China, India, Europe, Indonesia and Brazil can reach an agreement on Climate Change Control. A Carbon Cap and Trade system begun by the U.S. combined with trade sanctions against China, India and Brazil will likely get an international Climate Change Protocol passed by 2020.

If the U.S. does make progress in reducing imported petroleum by 2020, it will be a miracle.

Can a Sustainable Economy support fossil fuel? Should I invest in the stable, fossil fuel blue chip stocks and get the dividend in my golden years? Can Big Oil and Big Coal and Clean Energy synergize and create new solutions?

Yes, it is survival and adaptation of the fittest.

Same happens every day in Big Computer and Big Media Business. IBM buys Apple and you get an iPhone and iPad, and think coolio. Xerox, Microsoft, Intel and EMC buy local start-ups every day around the world. This is great news for Clean Energy and even New Media.

Small, innovative firms do not have the overhead to penetrate market barriers. Some go out of business, some succeed on their own and others are bought by the big power players in the industry they hope to supplant or supplement.

This is a business decision. Big Oil is already making Clean Technology acquisitions every day. Big Coal should follow suit and diversify in self interest. Markets change every day and every year.

Fossil fuel is finite. Clean and Renewable Energy holds great, unknown (and still testing) hope for petroleum import independence. It will boost the national economy by keeping Energy dollars domestic world wide.

By definition, fossil fuel is unsustainable. Some day it will get exhausted. It will not get regulated out of business. We know fossil fuel's harm and its benefit. It is the risk we accept freely. But for how long and who really knows?

New technology always faces challenges. Hidden by-products unknown, external social costs, job loss for those in sectors no longer needed. Sad but that is capitalism, half win half lose, every day.

Solar and wind look to become independent, big regional players in the Energy thirsty, growing Emerging Economies of the world, skipping the messy industrial phase.

China skipped telephone poles and millions of miles of wire infrastructure going straight to wireless. Africa and the rest of the world will follow suit.

There is always a fit between the old dog and the new kid, learning, nurturing, supporting, profiting and passing at different phases of growth. It is organic like our Carbon parent.

The world faces many challenges. We face mobility needs, economic vitality and development, a protected and safe environment done cost effectively but with laws enforced. We also require personal and national security. These can coexist.

The East and the West have grown closer, political philosophy recognized for is faddish importance, religion as a diverse and independently held personal faith that thrives in private and public with no threats to others every where in the world.

This is natural and organic. We are human. Our mutual enlightened self interest and sense of community responsibility, to town, country and now too the world.

The fortunate must help the less fortunate. The less fortunate must not threaten the fortunate. The world will come to accept in reciprocity the mutual growth, security and stability at the local, national and global level.

Rosy lenses. Realistic goals. Workable solutions. Maybe not even in our grand children's life time. But if we dare to dream and keep up the good work, we can thrive safely and sustainably with no fear or denial of our past or connected future.

So Clean Energy? BP bought you. Are you going to quit? There are dirty hands on my paycheck now. That is the question. No answers ever offered by me, just overlapping interests always hard to compromise on when you feel your answer is better and right. We know math. We know science. There are many ways to find the same answer. It is all in the proof.

There are a few simple things you can do reduce to use of gasoline and diesel. Most all of us own cars powered by gasoline. Follow your owners guide maintenance schedule. They are all online now. Maintenance can be costly in the short run, but you will keep your vehicle longer and less likely have an unplanned break down on the road.

Get the engine tuned at the manufacture suggested intervals. Change the oil religiously according to the specifications. Try not to cut corners for a few bucks. A tune up restores your mileage per gallon to near builder's specification. It is a Carbon reduction. Without the maintenance Carbon emissions would be higher. This is similar to replacing old drafty windows with new triple pain Argon filled windows or old, dilapidated insulation with new insulation. This is a change and with a bigger impact if you are improving with a newer, more efficient model.

Alternative Fuels gasoline and diesel with a lower Carbon content are available now. Waiting for a zero Carbon fuel that is an established market performer will take quite some time. If you are an early technology adopter and can afford it, you may want to buy a Natural Gas or Propane car, truck or SUV or even an electric car or hydrogen vehicle prototype.

The average person must wait for several years for new technology to get market tested and debugged. If you have an HDTV you can verify this because the audio is horribly inconsistent.

HDTV, compact fluorescent light bulbs, and high efficiency new air conditioners, central air and combined heating are fossil fuel reducers and Carbon savers that are market proven and getting better every new release. Good for the economy and jobs.

You could fuel switch. If you live in a state and town where you are allowed to choose your power generator source, chose the one you want. Power suppliers are mostly regional, national and international (Canada and Mexico).

A little toying with tools from your local utility gatekeeper electricity delivery company website can help you choose your personal goals, price point, town, county and state of generation and also the amount of Smog, Acid Rain and Carbon created and decide for yourself.

Nothing is ever perfect.

Nuclear has zero air impacts but huge local and regional water pollution issues since 90 of the 120 nuclear plants in the U.S. do not invest in plant safety or maintenance. These nuclear power plants are leaking radiation into groundwater, rivers, lakes and ponds that fish and plants live in and that many people drink and wash with.

Energy Conservation is effective to a smaller degree for the life style change. It only saves a few bucks on the electric bill. The behavioral benefit of going off grid to be with children is worth it.

The majority of the populace needs TV, the radio, the phone and the computer and printer nearly all waking hours. So do not let petroleum independence and Carbon Control make you think that Conservation is making more of a difference than new Energy Efficient Technology.

Do not confuse Energy Efficiency from new technology with Energy Conservation. You should not have to wear sweaters or ignore the need for air conditioning to lessen you oil or electric usage.

Buying new technology from your local supplier for pretty much everything in your home is good for the economy because it provides employment for every sector of the production and distribution cycle.

Your choices of power and maintenance do make a difference. This will not be earth shattering, maybe save you baby sitting money once a month. But with gasoline prices predicted to surpass $5.00 a gallon in 2011 due numerous market and non-market reasons, soon you will be extolling the Energy Conservation activists.

As a veteran strategic planner of the Clean Air Act of 1990 I can say clearly that petroleum receives an A for effort in cleaning itself up in the transportation sector, where it is predominantly used.

Petroleum in its diesel form heats homes in the North during the winter months. Diesel fuel is also still used to fire electric power generating stations but less so as the natural gas supply has become more plentiful. Natural gas is also much cheaper than diesel or gasoline. This is likely to remain the case for the distant future.

Petroleum, specifically gasoline for transportation, but even diesel for transportation, is the most highly regulated fossil fuel in the United States and Europe. The remainder of the world is wise to copy the gasoline and diesel regulations of the United States Environmental Protection Agency for transportation use in cars, trucks, buses and locomotives.

In the early 1980's Lead (Pb), the brain damaging poison still found in pottery coloring and the poison that deranged the minds of Roman Civilization through its Lead pipe water system, was removed from gasoline for transportation. Then the volatility or Reid Vapor Pressure of transportation gasoline was lowered to ease the evaporative pollution of parked cars. Yes, your car today pollutes almost as much when parked as when driven.

Since the 1990 Clean Air Act transportation gasoline has been re-engineered or reformulated several times. Diesel transportation fuel is now reformulated also for its specific nature.

Time and time again transportation gasoline and diesel have faced complex U.S EPA regulation. Law suits were rare and barely caught the national press eyes as most were negotiated away by expert engineers in industry and government.

Now transportation gasoline is facing new Corporate Average Fuel Economy (CAFÉ) standards for the first time since the late 1970's and is readily meeting them cost effectively.

Every time petroleum has faced stiff Federal regulation transportation petroleum has met the challenge. It has done so often very cost effectively for mere pennies a gallon.

This is in sharp contrast to the more volatile speculative pricing bubbles gasoline and diesel put the economy through in recent years. In fact it is this price volatility and not a Carbon Tax in any form that spells the most trouble for the future of transportation petroleum.

Not to be forgotten or ignored, the automobile, trucking, bus, and locomotive manufacturing companies worked closely with the petroleum industry to build engines that could run on the evolving types of gasoline and diesel over the past 50 years.

Not only did the transportation engine companies meet petroleum regulations they provided several types of vehicles that could operate on natural gas, propane, ethanol, methanol, electric batteries, hydrogen, and even celluostic biomass and Proton Exchange Membrane fuel cells with zero government subsidy.

In fact the decline of the United States automobile industry is directly tied to the Federal Government.

Transportation petroleum warranted regulation and Alternative Fuels were necessary. Government subsidy for the industry that provides one out of three jobs in the economy was and still is warranted.

But the Federal Government left the power generating plants operated on un-reformulated coal and diesel untouched by regulation. The utilities and electric power generating plants sought delay and resolution in the Court system, always losing to EPA, but effectively remaining unregulated for decades.

The petroleum and transportation engine industries were much quicker to drop lawsuits and leave the negotiating to engineers instead of lawyers.

Transportation petroleum with 70 percent imported from unstable foreign war torn regions of the world will unfortunately remain the primary transportation fuel as gasoline and diesel far much longer than coal will remain as a viable Energy source. The alternatives to coal are clear, numerous and easy. The alternatives to transportation gasoline and diesel are not market ready, clear, cheap, and still largely in small start up research and development laboratories.

It is possible petroleum may become synthetic and Renewable as a zero Criteria and Carbon Pollutant transportation fuel. The petroleum industy has consistently met environmental regulation and fuel economy improvements on schedule and under budget for decades.

Carbon Dioxide equivalent (CO2e) Greenhouse Gas emissions requirements will likely be achieved with exemplary engineering, cost, safety and performance improvements from the petroleum and transportation engine industries today and 100 years from now.

It is simply cheaper to solve the Carbon problem with a known equation and known fuel like petroleum than to even switch the fleet to natural gas vehicles. Americans love petroleum. The petroleum industry seems to always find more reserves of petroleum in almost every region of the world. The petroleum industry has the skilled workers, engineers, marketing and most importantly abundant supply of neighborhood distribution stations in every zip code in the United States to remain in the market for generations to come.

What will ultimately spell shrinking market share for gasoline and diesel as a transportation fuel? The free market of world capitalism has already spelled out the fate and doom of petroleum as a viable transportation fuel.

The death of petroleum is price volatility and price swings that have nothing to do with supply and demand but more tied to speculation, foreign government unpredictability, and investor and banking pressure on the petroleum industry to put stock price and profitability before safety and sound engineering. Poor management by BP has already tarnished the entire petroleum industry.

While viable for hundreds of years to come as an effective near zero Carbon and Criteria Pollutant transportation fuel, petroleum faces strong price competition. Due to the tremendous amount of known knowledge and engineering about the chemistry of petroleum it is possible to engineer away its environmental pollution.

But it is the continued poor management of the petroleum industry that will kill itself. Coupled with sooner than Big Oil realizes stiffer competition from Renewable and Alternative Clean Transportation Fuels that are much cheaper at the gasoline station, petroleum as a major market share transportation fuel grows increasingly less likely with each and every petroleum environmental catastrophe and price swing.

Nuclear Control Strategies

Nuclear Energy is less political now than it was in the past. It is still a political problem for engineers and politicians. The last nuclear power plant licensed in the United States was the Seabrook Station some four decades ago. President Obama has thrown his support behind building new nuclear power plants in the United States by providing financing for several proposed nuclear power plants on the drawing board.

The question remains though, does the private investment market want to invest in the remainder of nuclear power plants financing needs to get even one nuclear power plant built in the United States?

The answer appears to be no. The private banking industry is not willing finance new nuclear power plants.

Other parts of the world take a different view on the role of nuclear power in the future. Europe has a large supply of nuclear power. Yet when a vote was taken of the German populace by Chancellor Merkel about whether to continue investing in the existing nuclear power plants in Germany, the public vote was a responding No.

Despite the strong current public opinion against nuclear power in its existing facilities never mind construction of new plants, the German Government went ahead and voted to continue investing in nuclear power.

China, an authoritarian state that is drifting slowly towards more openness takes a vastly different view on nuclear power. With the help of dozens of United States engineering and construction firms, most notably Shaw Group, China is constructing dozens of new large nuclear power plants in the next decade.

China, faced with some of the worst traditional air quality pollution in the world from its coal power plants and increasingly automobile oriented populace, sees nuclear power as a key part of their urbanization and modernization plans well into the next generation.

China is a state run economy with leaders and bankers that do not have to face public opinion in the polls or shareholder proxy votes.

This makes it easier for China to forge ahead with bold, aggressive plans for vast nuclear power plant expansion. A similar situation is highly unlikely in the United States and Europe and less likely in other parts of the mostly capitalist world.

Nuclear power is not Renewable like wind, solar or hydro. Because such small quantities of fuel are needed it could well be a major force in the Renewable Energy world economy for hundreds of years to come. This is especially true if there is a fusion power breakthrough.

The only problem with nuclear Energy is its waste. Currently, arms treaties prevent the re-use of spent nuclear fuel. The Yucca Mountain National Nuclear Waste Repository in the planning phase for nearly half a century appears to be finally dead. There is no national plan on how or where to store the long lasting nuclear waste in the U.S.

Making matters worse, 90 of the 120 or so existing nuclear power plants in the United States are leaking dangerous levels of tritium. This and other cancer causing nuclear waste products are in our local public drinking water supplies in nearly every state in the country.

If the United States can not even fund the maintenance of existing nuclear power plants, how likely is the public and private investor to invest in new nuclear power plants?

Nuclear power in the United States has attracted founding people of the Environmental Movement from the 1970's to support it.

They see Climate Change and Carbon Control as more important than the long lasting carcinogenic reality of nuclear waste. Coal pollution kills people now. Nuclear pollution kills your great-great grandchildren.

The next time a new nuclear power plant tries to get a new operating permit in the United States it will fail. The long lasting irresponsible use of a fuel with known hazards and costs will outweigh any perceived benefit.

We are not willing to pay for spent nuclear waste now or likely in the future. Nuclear power in the United States and the rest of the Democratic, capitalist world will face rejection from the public at large and from investors and shareholders.

President Obama announced $8.3 billion in Federal loan guarantees to two proposed nuclear reactors in Georgia in the early spring of 2010. The move may disturb some Democrats and environmentalists, but is a good gesture at seizing a key Republican item.

President Bush authorized about $18 billion in guarantees in his 2005 Energy Bill. Obama hopes to triple that to roughly $55 billion. Department of Energy Secretary Stephen Chu says the $55 billion could help lower construction and financing cost for approximately seven to 10 new reactors.

Vermont Yankee nuclear power station is suffering radiation leaks reaching the adjacent Connecticut River, in Southern Vermont. Governor Patrick (D-MA) called for a safety check of the Plymouth Pilgrim nuclear plant that was deemed safe.

Obama's announcement came as an Energy Bill and Climate Change Control were stalled in the Senate by Republicans and Moderate Coal State Democrats from the Midwest, like Evan Byah from Ohio.

Senator Lindsey Graham (R-SC) said he hoped the announcement would revive talks with skeptics in the Senate who oppose passing the bill and putting a price on Carbon. It did not. Senator Graham pulled his support of Climate Change Control legislation in 2010, in favor of immigration reform.

Natural Gas Control Strategies

In the transition to Clean Energy there are bumps along the road for natural gas. Around the country states and local governments are mandating local utilities to purchase more Renewable Energy power.

In La Plata County, Colorado solar is the Clean Energy of choice. Tax credits, rebates, abundant sunlight and Climate Change Control desires are slowly increasing the role of solar in Colorado.

La Plata did a Greenhouse Gas inventory and found its biggest source was Methane from natural gas pumps and associated equipment that bountifully fills the landscape. It also pays the county with tax revenue.

Natural gas faces a double whammy. Reducing GHGs from its production equipment is increasingly expensive.

When the nation's biggest electricity buyers local utilities pay the full external impact cost of Carbon coal pollution, natural gas prices will go up.

Being able to put solar systems on homes allows removing from the grid, like Energy Conservation. We will need natural gas. The industry wants to be embraced wind, solar and Renewable Energy, which is not likely.

In 2009 the natural gas industry formed a new lobbying group, America's Natural Gas Alliance. The group stresses natural gas as cleaner than coal and the transition fuel to a zero Carbon Energy economy.

Political leaders in Energy rich states with numerous fuel types have to avoid getting pinched by the competing fuels.

The Waxman-Markey bill passed in June 2009 largely shuts natural gas out of Federal subsidies. Senator Mark Udahl (D-CO) aimed ensure the final bill on Climate Change does not prefer some home state Energy sources over others.

The natural gas industry pays half the property taxes in La Plata County. It provides hundreds of jobs. Around 6,000 residents with mineral rights on their land get a monthly royalty check from oil and natural gas companies. Solar can not do that. Solar costs money.

Many transit agencies are switching to low Carbon bus and rail fleets. Many issues need addressing in evaluating a new low Carbon fuel for transit systems.

Transit agencies already using diesel, assessing dual fuel, low sulfur – electric battery powered systems with a well equipped and maintained particulate matter trap cuts down on cost of a different traditional fuel filling.

All-electric battery and overhead catenary powered buses and locomotives are available on the market where feasible, depending on cost and neighborhood impact.

Many transit fleets switched to compressed, liquefied natural or propane gas in the 1990's due to asthma impacts and air pollution. These three fuels have lower innate traditional emission profiles than diesel.

The half life of Methane in the atmosphere is months compared to Carbon from diesel dual fuel buses or locomotives. Carbon Dioxide lasts hundreds of years in the atmosphere.

School bus fleets are also assessing low Carbon fuels that are safe, reliable and cost-effective in many regions. Natural gas is an urban fuel, propane gas a rural fuel.

Celluostic biofuels from man made sources are a zero Carbon-equivalent transit fuel. It is often locally made and distributed, instead of imported.

Battery packs for large buses and trucks are available with improved battery time operation and quicker charging times, depending on the city.

Regional inter-city passenger rail service in China is demonstrating low Carbon Class 1 (long haul) locomotive engines.

Class 2 (short haul) and Class 3 (switching yard locomotives) are most easily and cost effectively converted low Carbon engine types.

Electric overhead catenary transit is good for regional commuter rail. When the New York City commuter rail system electric in the early 1990's, an unintended consequence occurred. Trip times improved. Diesels are difficult to slow or speed up. Diesel engines also experience performance deterioration and reduced speeds on inclines and declines.

Clean Fuel is traditional fuel, usually fossil based that will play a role in bridging the world economy to a Carbon Zero Renewable Energy future of wind, solar, biomass, hydrogen, hydro and tidal Energy.

Liquefied natural gas (LNG) is a perfect low Carbon, short lived Carbon Dioxide equivalent fuel for heavy duty engines involved in long haul fright transport across long distances. Liquefied natural gas is compressed natural gas in a frozen state allowing much greater fuel storage aboard heavy duty freight vehicles allowing comparable refueling with diesel.

Liquefied natural gas (LNG) is transformed to compressed natural gas before it is combusted by the heavy-duty engine. Liquefied natural gas' main emission is Methane (CH_4) and to a lesser extent Carbon Monoxide (CO), Volatile Organic Compounds (VOC) and a small amount of Nitrogen Oxides (NOx). Particulate Matter (PM) is virtually zero.

Natural gas is projected to remain $1.00 cheaper than diesel for decades to come thanks to the enormous North American supply of natural gas.

Liquefied natural gas is safe, efficient, inexpensive, environmentally and greenhouse gas friendly compared to diesel, the primary fuel of inter-state freight.

Liquefied natural gas is perfect for long haul Class 1 rail, inter-state semi-tractor trailer trucking and shipping. All three of these major freight carriers have something airplanes and light duty cars and trucks do not have, plenty of power and storage space for the liquefied natural gas tanks.

Liquefied natural gas tanks come in double and triple shell bodies that are extremely safe from external puncture in an accident or incident.

Rail would be easy to fuel as locomotives run fixed routes across the nation. This is also true for shipping in national and international waters with ports that could easily fuel liquefied natural gas shipping.

Interstate trucking could focus first on the major East-West and North South Interstate Highway System, eventually filling in the feeder highways.

The economics of natural gas compared to diesel and the closest Renewable Energy counterparts of biomass, hydrogen and electric are clear.

Liquefied and compressed natural gas vehicles are market ready with no incremental purchase price.

The fuel price savings of natural gas more than offset the new costs for building out the fueling infrastructure. Liquefied natural gas as a bridge in a low Carbon future now to an eventual long term replacement fuel is clear.

Once the market shakes out the technical bugs and economic externalities of Renewable transportation fuel liquefied natural gas may be even cleaner through new technology breakthroughs.

Until the day an affordable market ready Renewable transportation fuel has been tested in the market, liquefied and compressed natural gas are clear low Carbon transportation fuel winners.

Natural gas is the clear winner in the transition to a Renewable Energy economy. Natural gas is important to both power generating plants and transportation for well into the foreseeable future.

Natural gas' innate chemical formulation of primarily Methane (CH_4) and zero Carbon Dioxide (CO_2) give it the great attribute as a bridge to a Carbon zero economy. Methane (CH_4), also a greenhouse gas (GHG) emission, remains in the atmosphere for mere months compared to hundreds of years for Carbon Dioxide (CO_2). It is impossible to deny that natural gas is vital to world economy of the next 100 years.

Natural gas is also plentiful on every continent in the world. New deposits of natural gas are found monthly in North America. Extraction of natural gas from plentiful U.S. and Canadian shale deposits have greatly increased the long term supply of clean natural gas well into the next century.

The enormous increase in the market ready supply of natural gas in the last few years has vastly decreased its cost to consumers. Natural gas has always been cheaper than petroleum and coal.

Natural gas is easier to extract from underground than coal and petroleum. It does not require expensive refining or reformulation. Natural gas requires very little air quality pollution control technology when compared to coal or petroleum.

Natural gas is expected to remain at least $1 below the cost of petroleum for years to come. The price of petroleum is more likely to increase over the next several years than decrease or even stabilize around $4 per gallon.

Many analysts see $5 to $6 a gallon gasoline in the near future. Natural gas is cheap to extract, cheap to process, cheap to distribute, cheap to use and is very clean on both Criteria and Greenhouse Gas pollutants.

Many states around the country have locally deregulated their electric utilities. This allows consumers to purchase the electric generating fuel source from any supplier the local distribution company or gatekeeper uses.

As more and more consumers have become educated about Carbon pollution, many have switched to new natural gas power plants as their electricity generating power source.

Many states now have a Renewable Fuel mandate to increase the market share of Renewable Energy.

The governments have often failed in mandating which fuel or fuel type should be the market winner. Renewable Energy remains expensive.

Wind and solar power will never be able to compete in the Energy market without government mandates to use more Renewable Energy. The governments failed in the past to follow through on mandates for Alternative Fuel types for transportation.

It is just as unlikely that the mandates for utilities to purchase Renewable Energy will remain in place. Once consumers realize how much more expensive Renewable Energy is the mandates on utilities will be relaxed.

Natural gas wins as an electricity power generating source, as a home heating Energy source and as a transportation fuel. Natural gas is clean, cheap and domestic.

Wind as a Control Strategy

Should wind farms be sited in the Great Plains or on the East Coast? A debate in the Senate in 2009 was whether new wind Energy should be developed in the Great Plains states where it is abundant versus local wind farms on the East Coast, closer to sources of need.

At the heart of the debate is whether to construct massive new transmissions lines from the West to the East to carry solar and wind power. The current grid is already maxed out to capacity.

East Coast Governors appear to have won the debate. They argued it is better to let the market decide where wind farms should be built. They argued more jobs would be created on the East Coast if local and regional wind farms were constructed.

Obama supports a new national transmission grid to support Southwestern solar and Great Plains wind farms Clean Energy transmission and distribution to the East Coast.

The West Coast has always relied on long distance power generation and distribution. The East Coast with the Energy need, population, and higher unemployment wants the new wind Energy jobs created locally. In the Great Plains there are fewer qualified workers. It is hard to find urban amenities.

There should be a mix of both local and long distance wind Energy production. If the new transmission lines are built coal power will be carried on it too, possibly raising Carbon emissions.

It appears in the near term local wind farms are favored in the Climate Change bill. But expect years of legal battles between wind Energy companies and scenic preservationists.

The need to transition to a reduced Carbon Energy supply outweighs the negatives of East Coast wind farms.

Wind power was up 40 percent in 2009 due to the Stimulus. The U.S. now gets two percent of all power from wind. The wind sector grew at a blistering 40 percent in 2009. About 10,000 Megawatts of wind Energy were added to the power grid in 2009. The Great Recession has lowered growth of new orders for large wind turbines.

The growth in wind power electricity capacity equaled natural gas. These zero and very low Carbon Energy sauces made up 80 percent of new U.S. power capacity added in 2009. Wind power growth is up 700 percent since 2000. 85,000 people are employed in the Renewable Energy sector in the U.S. Half of the new installations of wind power in the U.S. are domestically produced. This is more than double just a few short years ago.

In contrast, Europe gets five percent of its power supply from wind. Europe has set a 20 percent wind power market share for 2020. China invested $15 billion in 2009 in wind generation. This more than doubles their wind power capacity by the end of 2010. China plans aggressive funding of wind power growth through 2020 at least.

West-East transmission grid expansion to promote the explosive wind Energy growth in the Great Plains and Mountain states will not disapper. Many fear it will be used for new coal and natural gas projects. A national power transmission grid investment meets with stiff opposition from East Coast politicians who want local wind power for local jobs.

Most East Coast wind projects are highly political, opposed by scenic view naturalists, tourism officials, Native Americans and real estate lobbyists who do not want to see massive tracts of industrial wind turbines on scarce, expensive, high value real estate

It is projected by 2020 about 180,000 Megawatts of wind power will be on the market. This is about five times more than today. Wind power energy lights and heats 10 million homes in the U.S. Texas is the largest wind producer with 10,000 Megawatts, followed by a distance Iowa with 300,000 Megawatts and then California, Washington and Minnesota.

Solar as a Control Strategy

On the July 4, 2010 media address President Barack Obama unveiled and announced a massive $2 billion direct Federal investment to two solar plants. The funds are from the Wall Street firms that complied with the winter 2009 Obama Order to repay the short term financial market rescue public taxpayer dollars as soon as possible at high interest.

"We're going to keep competing aggressively to make sure the jobs and industries of the future are taking root right here in America," President Obama was quoted.

Abengoa Solar is one of two Energy Technology solar firms awarded the investment, constructing the world's biggest solar farm in Arizona. 1,600 constructions jobs were created. Abound Solar Manufacturing is building solar farms in Indiana and Colorado. Abound Solar hired 2,000 workers.

Energy Technology is a multi-trillion dollar world wide economic sector found in every G-20 country. Energy Technology employees engineers, planners, web managers, top notch economists, project controllers and executives in construction and operations.

Energy Technology is fast replacing and surpassing industrial age coal, oil and heavy polluting industries like steel, cement and glass. These industries are still vital but cheaper, cleaner and efficient.

Energy Technology is challenging researchers from grade school students in India and Indonesia to economists in South Africa, China, and Brazil to create local, Carbon free, sustainable, cheap Energy sources to develop and maintain the world's power needs well into future generations for a more regional, stable world economy.

Electric Batteries and Hydrogen for Vehicles

The search continues for cost effective super electric batteries. The race is on to build a smaller, lighter, more powerful battery for cars, windmills, and solar panels. In the Renewable Energy field, efficient batteries absorb extra Energy when output is great and discharge the Energy when the wind slows or clouds appear.

The price of batteries is a key market barrier. Car and grid batteries are thousands of times the size of cell phone batteries. Research on lithium ion batteries is at the forefront of the market.

Appearing in the early 1990's, lithium ion rapidly displaced nickel metal hydride batteries in portable electronics. Lithium, the third lightest element in the periodic table, allows for greater Energy density.

Lithium will transport a car one mile at half the weight of a nickel metal hydride battery and 1/6th that of a lead battery. High tech battery production is centered in Asia in Japan, China, Taiwan and South Korea, the laptop manufacturing center of the world.

Battery technology also faces many technical hurdles beyond cost. Batteries must store great Energy in a small, light package creating high Energy density. They must also rapidly cycle uptake and release of electricity for good power density for thousands of cycles. Each of these goals has been met, but not in one product.

In 1991 the Advanced Battery Consortium set a cost goal of $150 per kilowatt-hour (kw-hr). A kilowatt-hour costs $0.10 and moves a car 3/4ths of a mile. Almost 20 years after, the cost is between $750 and $7,000 per kw-hr. The cost for lead-acid batteries is only $100 per kw-hr, but they are too heavy and not durable. The new Department of Energy goal is $500 by 2012.

Obama is handing out grants under the Stimulus Bill totaling $2 billion. Obama's plan gives tax credits as much as $7,500 for consumers who purchase plug-in hybrid cars, a mix of gasoline and electric power. "This investment will not only reduce our dependence on foreign oil, it will put Americans back to work," Obama said. "It positions American manufacturers on the cutting edge of innovation and solving our Energy problems."

In February, Consumer Reports reviewed a jazzed up Prius conversion to a plug-in. At $11,000 the Prius had a five kw-hr battery installed at a Toyota dealership in Massachusetts. The results: 67 miles per gallon and a 35 percent increase over the original gasoline powered Prius.

Some experts think nano technology looks promising. Nano materials have very large surface areas in small, light packages. Batteries function by chemical reactions occurring on surface structures. Batteries work by particles moving through electrolytes, an inefficient process.

Nano materials have problems, becoming clogged and degenerating after a couple dozen cycles. Solving the problem may mean cheaper, lighter batteries.

The plug-in Chevy Volt that debuted in late 2010 has a 16 kw-hr battery and travels 40 miles per charge. The Volt's battery package ranges in cost from $10,000 to $16,000, not including the required batteries' support systems. Technological and market barriers will remain for years for super efficient, cost effective batteries.

In mid 2009 the U.S. Department of Energy ended research into hydrogen fuel cells.
The 2010 Budget for the Energy Department ended research into hydrogen based fuel cell technology.

Massive manufacturing facilities had been planned but have been scrapped. Some funding will remain to continue to find better battery solutions. The hydrogen industry will no doubt view this as a serious set back.

Even President Bush said in his last term that the car your child will drive will be a hydrogen fuel cell car. Secretary Chu of the U.S. Department of Energy does not feel the technology is market ready for prime time deployment.

I have never seen a hydrogen vehicle on the road, never seen a hydrogen refueling station for fleets or private vehicles. Electric cars appear to be in the same boat, as do other new technologies that do not have the might of the natural gas or propane industries. Even though natural gas and propane are Carbon based fuels, they emit far less Carbon than gasoline or diesel.

According to the Massachusetts Executive Office of Energy and Environmental Affairs the average New England household emits 2,137 pounds of Carbon a month, while a natural gas home emits 359 pounds of Carbon equivalent emissions, mostly in the form of Methane. Propane homes only emit 65 pounds of Carbon equivalent emissions per month.

While hydrogen may be one of the fuels of the future it is not the fuel of today. The same thing applies for electric cars.
Natural gas and propane have 20 years of vehicle technology development and improvement and have been market ready for decades.

The refueling infrastructure for CNG and propane is not comparable to diesel. It is slowly expanding, even if the pace does not match government hopes 10 years ago. How often do you see a diesel station?

Hydrogen and electric cars are going after the private passenger car market fueled by gasoline. This is a much more difficult market to tackle without government mandates than the fleet/truck sector has received. The modest government incentives to switch to Alternative fuels since the early 1990's are helpful but not enough. Alternative fuels for fleets continue to receive very modest grants under the Stimulus bill.

The Congress gave SUV owners $4,500 cash back incentive to buy a newer more fuel efficient SUV even if the gas mileage improvement was two miles per gallon.

Gasoline and diesel will long remain dominant fuel sources in the transportation sector no matter how ambitious the goals government set for conventional fuel displacement of imported petroleum.

I see fuel diversification of the transportation fuel sector in the next half century. If we can cut into it by 50 percent that itself would be a victory, even if it takes until 2050. There is a place for hydrogen and electric and non-corn based celluostic ethanol vehicles in the future. It will vary by region and market and their will be no clear winner.

The only loser in this game is that it makes much more expensive for vehicle manufactures to deploy market ready Clean Fuel vehicles for so many fuels the public will buy. Federal assistance in the next half century to vehicle makers is warranted especially as they struggle through the Great Recession and try to maintain market share in the global vehicle market.

So, hydrogen, electric and ethanol, welcome to the table, but CNG and propane have been doing battle with diesel and gasoline for 20 years.

Congress did restore funding for the hydrogen car. Energy Secretary Chu cut out funding from his budget request for research into the hydrogen car. Congress restored full funding.

The hydrogen industry requires over $55 billion in additional funding in the next 15 years. The appropriation was only $187 million.

Hydrogen cars face more hurdles than other Clean Energy fuels. Cost and practicality of building hydrogen cars and the extensive fueling infrastructure of stations are serious impediments.

Serious technical problems with the manner hydrogen is derived, stored and distributed are immense market barriers compared to natural gas and propane cars and trucks. Fuel cell technology is poorly developed, expensive and limited in operation duration and range.

Cost is the major market barrier.
GM, Toyota and Daimler fund research in hydrogen cars, but no one is even thinking about telling vehicle showroom price. Test cars cost more than $300,000, but mostly because they are custom builds and not mass produced.

There are fewer than 12 fueling stations for hydrogen cars in the U.S. Hydrogen fueling station cost is a staggering $2 million almost 10 times more than gasoline, diesel, natural gas and propane fueling stations.

GM has invested $1.5 billion in hydrogen car research. GM says the weight of the engine is greatly less than it was a few years ago. The use of expensive precious metals has been cut in half. Hydrogen cars could be in your neighborhood showroom in 2015.

Energy Efficiency and Conservation

Can Energy Efficiency work? After 35 years since the first Arab Oil Embargo, two reports claim the U.S. can achieve its 2020 Carbon reductions with no Energy growth cut backs all through improved Energy Efficiency.

The McKinsey report and the U.S Academy of Sciences, National Research Council each claim rosy Energy Efficiency gains in the next decade. McKinsey claims a 23 percent Energy reduction by 2020 while the National Research Council projects a 15 percent reduction by 2020.

40 years ago it was wind and solar after the first Energy Crisis in the 1970's. Then in the 1990's the electric car raised hopes, along with Alternative fuels.

What does the U.S. government have to show for it? Not much. While Energy Efficiency in homes and commercial buildings has proceeded for the past four decades, at best they staved off demand in Energy growth and new power plant construction.

This is a big accomplishment. Energy Efficiency should be praised. But Energy Conservation and Efficiency will never actually reduce Energy consumption, just simply offset growth.

The Stimulus Bill contained about $12 billion for Energy Efficiency. It went to state and local governments. Some money went to industry.

A Watertown, Massachusetts battery maker received $250 million from the U.S. Department of Energy as did General Motors mostly for its electric Volt.

McKinsey claimed the U.S. would need to invest $50 billion in Energy Efficiency annually over the next decade, clearly unreasonable, despite potential gains.

After a generation of Green Technology, most companies remain small and in start up mode. Many never take off, achieving profitability. Many go out of business. Some start ups are gobbled up by bigger industry players, as they try to evolve to a changing world market. One should never underestimate the power of big industry.

Is Energy Technology the next big thing?
Some call it Clean Technology. Obama calls it Green Technology. A new term is popping up: Energy Technology. It is already changing the landscape of the world economy.

The leading country in Energy Technology is China. China missed the Industrial Revolution and was still slumbering during this generation's Information Technology Revolution. But China is already leading the world in the Energy Technology Revolution, focused largely on its strong manufacturing base and a cheap, abundant labor.

As China is transforming in the largest rural to urban migration ever in the worlds, China is leading the way in Energy Technology and Climate Change Control, even without mandatory caps on its Carbon growth and limits.

China announced the largest yet solar project, a two gigawatt project using California based eSolar technology. In contrast eSolar has a 92 megawatt solar project in New Mexico.

China is ready to begin construction on its new two gigawatt solar facility while eSolar is still working its way through the Department of Energy's grant application bureaucracy.

As dozens of new solar panel makers have spawned in China in 2009 the cost of their solar power has plummeted from 59 cents a kilowatt hour to just 16 cents.

China tested the world's fastest high-speed rail system at 217 mph from Wuhan to Guangzhou.

China is almost done with construction on a bullet train from Beijing to Shanghai, covering the 700 miles in less than five hours in contrast with the current 12 hour trip, at a cost of $23.5 billion. In the U.S. the comparable distance is New York to Chicago, a trip that takes 18 hours on Amtrak.

China is also the world leader in new nuclear Energy power plant construction. China has plans to build 50 new nuclear plants by 2020 while worldwide only 15 new nuclear plants may be built.

China is making clean Energy cheaper for itself and the world too.

The U.S. should lead the world in Energy research, invention and development as well as servicing and maintaining the new Clean Technologies. China will remain the as the world manufacturing center of Clean Technology.

The U.S. needs to put a price on Carbon to secure our Energy, security and economic power while enhancing our short and long term environmental protection and quality. China may have built the world's fastest bullet train but the U.S. is steering the course of the Energy Technology Revolution.

It is now possible to give control of your appliances to your utility through the new Smart Grid. Peak Energy demand on hot summer days in the U.S. can cause rolling brown outs as utilities try to avoid large scale blackouts.

General Electric's Smart Grid technology conserves
Energy by switching off appliances in the home when
electricity is expensive during peak demand and
wind and solar power are at prime generation rates.

Smart Grid technology takes voluntary measures out
of the consumer's choice. It automates them saving
the customer money and the utility valuable peak
spreading Energy. There were over 70 Smart Grid
demonstration projects around the country in 2009
including one launched in Maui, the first ever resort
to try it, and also in Seattle and Boulder, Colorado.

There are opportunities to improve what is using
power and to shut off power when we are not
around. Most consumers are not really aware of how
much Energy they're using at any time of day.

By knowing when power costs most, consumers
could adjust the air conditioning and refrigerator
temperature or delay using the dishwasher or
laundry until the power demand peak drops. This
saves money on the home utility bill and saves the
grid capacity allowing it to absorb more wind and
solar power during the day when it is at its peak
output.

Are you thinking of investing in a Clean Technology
company? Choose wisely. The market barriers
stacked against them are immense from traditional
fuels. Traditional fuels can invest more in cleaning
up their old fuel than a start-up can gain from venture
capital.

Expect any money you invest in Clean Technology to be a complete loss. Do not bet big. Do not expect a rapid return. Expect to lose your money. If you can not accept this risk you should not invest in Clean Technology.

Why? Because very few of the technologies will make it market. The Obama Plan was temporary. Government funding has never been enough to keep a start up in business. All the government can do is benchmark market ready technologies. The government is not and should not be a market incubator.

This is your job. You will lose your money. But venture capital is the only way Clean Technology will make it in the next decade. Government mandates may help, but they do not take affect for years. Never rely on the government to create a market. It only destroys or over regulates an industry. The government is a watch dog and a very poor one at that.

The job of creating the new, advanced, Clean Technologies of tomorrow is yours and your fellow investors. Follow traditional markers and ratings. Do not go by company statements. Check for membership in a national association, the web page, age of industry group, members, focus and regional chapters.

Lobbying and marketing is important. There are hundreds of Clean Technology companies out there. How do you know which ones are real and which ones are scams. Treat it as a traditional high risk junk bond.

Do not have false expectations of Clean Technology making you rich or saving the world.

Green Technology jobs were shaping up in 2010. Everything from what you would expect, small Green Technology start ups to even large Wall Street brokerage houses and investment firms looking to make money off of Carbon Control, even without an international treaty or legislative passage of Cap and Trade in the U.S

Look to follow, comment, prognosticate and advise on legislative processes and regulatory promulgations. This is a lengthy process. I did not expect legislative action by Congress in an election year after the health care battle. Look for EPA to take the lead in 2011 after the mid-term elections. It will take EPA until 2012 or 2013 after the next Presidential election to fully implement a Carbon Control program.

EPA is not moving forth with a Cap and Trade scheme but their time proven Operating Permits under Title V of the Clean Air Act of 1990. Cap and Trade may emerge again at EPA a few years down the road so this will have to be watched closely.

In 2011 and 2012 MBA's with less air quality experience can look to investment and brokerage houses to provide advice to investors on the changes of marketability and profitability of Green Technology firms. You can support a family off these jobs, but do not expect to earn what a Street insider or investment banker or Goldman Sachs guy makes.

In 2009 President Obama announced investment in the United States Smart Grid. In Florida President Obama announced a $3.4 billion Federal investment in Smart Grid residential electric meters. Utilities are matching that with $4.7 billion of their own investment.

Smart Meters allow your home to communicate with your utility, so for instance, when power is out, the utility is automatically notified so you do not have to call. The President says the program will take place over the next three years, creating tens of thousands of new domestic jobs.

Smart Meter technology frees up space on the electric grid by monitoring power loads and avoiding rolling brown outs and major black outs. It also allows the electric grid to absorb more Renewable Energy from wind and solar during the day when those sources contribute most. It saves fossil fuel usage for the overnight hours when demand is lower. This saves the consumer money.

The indoor, residential Smart Meter visually displays the information from the Smart Meter on the outside of the home. It displays Energy usage in kilowatt hours and cost per day. It shows a comparison by time of day, day of week, week to week, month to month and year to year.

Instead of manually needing to turn unnecessary appliance use off during the day, you can schedule your Smart Meter to automate when you want to run your dishwasher, dryer, washing machine and other appliances.

President Obama unveiled in 2009 the final Carbon dioxide and gas mileage standards phasing in between 2012 and 2016.

The average fleet wide standard is 250 grams per mile of Carbon Dioxide and a 36 mile per gallon fuel economy. The new vehicle cost to you is less than $1,000. Annual payments on a five year loan will save you $160 a year. You will save $3,000 in gas costs.

A compact Honda Fit will create 200 grams per mile of Carbon with a fuel economy of 42 mpg. A mini-van like Toyota Sienna will create 300 grams per mile of Carbon with mileage around 28 mpg. A large pickup truck like a Chevy Silverado will make 350 grams per mile of Carbon with mileage around 25 mpg.

EPA's standard for nitrous oxide is 0.010 grams per mile. The standard for methane is 0.030 grams per mile

There are credits for meeting the standards earlier, introducing more Alternative fuels, reducing air conditioning leakage and advanced technologies. There are waivers for small volume luxury car makers.

Regional Greenhouse Gas Initiative (RGGI) adopts Alternative Fuel Rule. The RGGI adopted a Low Carbon Fuel Standard for the transportation sector. It may be applied to residential and commercial buildings sector, a breakthrough first.

Transportation accounts for 30 percent of Greenhouse Gas emissions in the heavily populated Northeast, one third of the nation's whole population.

This is similar to the Clean Air Act of 1990 Clean Fuel Fleet Standard which the Northeast chose to replace with Reformulated Gasoline for light duty vehicles instead of burdening the commercial sector with more regulation. The Energy Policy Act of 1992 called for government fleets to get off petroleum and utilities too, but DOE balked at passing the rules onto the private sector by the turn of the century. The Energy Policy Act had ambitious petroleum displacement goals but failed miserably to even reach a one percent displacement factor.

Market barriers remain for all Alternative Fuels. More costly vehicle purchase price, usually more costly fuels, patchworks of fueling station making it inconvenient to fuel and very modest government tax incentives. Safety, driving range and vehicle choice have improved vastly since the early '90's but still remain poor.

Petroleum is and will remain king without Cash for Green Cars program by the Federal and State Governments.

Yes, you, the consumer, are more affective than the government at controlling Carbon. Here are some simple, low cost things you can do. In many markets, local utility distribution companies are installing smart meters on homes, apartments and offices to help consumers monitor Energy usage. It allows voluntary control at home and or via the Internet.

You have to buy a reader for your living room so you can get the data. It tracks Energy usage by hour, day of week, month and time of year, year to year. Many companies have voluntary targets to help consumers reduce Energy consumption in their home by 10 percent per year. There is also a voluntary option to allow the company to turn off things like washer and dryers during hot summer days to avoid rolling brown and black outs.

Behavior modification does not achieve as much for Carbon reduction as buying new more efficient technology and getting and Energy audit. Audits are free from most local power suppliers but in some places you must pay a qualified professional. They can quantify how much money it will cost, the return on investment and breakeven and savings point in terms of dollars, years and Energy saved, well worth the time and cost.

Many utility markets are deregulated now and allow you to choose what type of source you want to generate your Energy. You can select an aging coal plant or new clean natural gas fired plant or growing Renewable Energy sources like solar, wind and hydro. The more of us who purchase Renewable Energy generating sources, the more private investment it will bring to sustain them.

So, you can do a great deal more than the government can to slowly change Carbon patterns. Real, myth, deadly, delayed, unknown or otherwise. It just is plain, common sense and saves money now.

Energy Efficiency and Conservation deserve more attention. What is an Energy audit? There are home and commercial Energy audits. Home Energy audits do reduce Carbon if more efficient heating, cooling or lighting systems replace old technology. There is improved insulation and replacement windows which will increase your energy savings. A home Energy audit is available through private auditors or maybe the utility.

A commercial Energy audit is where the big Carbon reductions occur. The same categories are here, heating cooling, lighting, water heaters and chillers. Equipment is usually not included. Some companies can achieve a 20 percent reduction in Energy use. For the residential user to a small business to a business with a large Energy usage the cost and Carbon savings are quite large.

When it comes to another kindred topic, Energy Efficiency, this is where equipment, building design, supply chain, production and distribution are assessed by internal or external professionals. A movement towards greater social responsibility by shareholders and corporations to lower Carbon output is taking place. Many businesses assess Energy usage and Carbon output in core business functions.

Many recognized professional registrations are now available to qualify people for various types of Energy Efficient, low Carbon impact design, production and transport professions. Software modules on the market as added components of existing enterprise wide accounting system are available.

This may be in response to a company goal or government requirement.

Energy auditing either residential or commercial is almost as 40 years old. Around to improve business forever, it was transferred to homes after Energy price spikes in 1970's. Oil over charge money from big petroleum from government fines funded residential Energy audit programs into the late 1980's.

Energy auditing, or Energy Efficiency, is now a self sustained private venture. Some regional utilities subsidize the cost to home owners and business for an Energy audit. In some regions the individual must pay for similar to any home inspection on the local market.

Energy auditing is a good career path for anybody young or old regardless of education. Most employers sponsor training. The job involves using your hands and a computer to calculate the cost benefit pay back time for a new heating system, insulation or replacement windows. Air conditioning and home appliances are also covered.

Commercial Energy audits can save business of all sizes enormous amounts of capital that can be used for any other non-core business expense from new products, to marketing to new hires, to simply making a profit to stay competitive. Energy efficiency in the building shell and utilities is supplemented by the appliances and machines operating off the power supply. Increasing the Energy efficiency of both the source power and uses by equipment in the plant or home create huge cost savings.

New electronic monitoring devices allow owners to monitor Energy use and cost and modify operations according to choice.

It reduces the new need for new power generation. In the four decades of Energy Conservation we have not reduced Energy use, simply offset its growth. This in itself is a huge success and made Energy Conservation a private self sustained business.

Climate Change Affect on Wildlife

Climate Change impact on wildlife and walruses. Coastal ocean ice is missing along American and Russian beaches in the Bering Straight. Nearly five decades after bouncing back from massive factory hunting, walruses are once again in trouble from man. The Chukchi Sea, the homeland of the Pacific walrus, was largely open water in the summer of 2009 and 2010, even though the ice melt was not as bad as the past few years.

Biologists from the U.S. Geological Survey published a document determining 131 dead walruses found at Icy Cape, Alaska on September 14, 2009 died from a stampede. Thousands of walruses were gathered on the beach, which is abnormal. In September 2009, the World Wildlife fund documented 20,000 walruses on the shore of Cape Schmidt, Russia. In 2007 the area saw several thousand walrus deaths from stampeding after tens of thousands of walruses crowded the shoreline.

Deadly stampedes with walruses have happened before.
In 1978 a similar event was reported. Reports from hunters in the Bering Sea also documented walrus stampedes in the 1900's. As the ice retreats the ocean becomes more open, the shoreline shrinks because of rising water levels and places more stress on the walrus.

For 15 million years' walruses have survived dramatic climatic swings. Extinction is not predicted for the walrus, especially if hunting remains regulated. Native hunters in Russia and the U.S. are permitted to legally kill thousands of walruses each year.

The arctic ice flows provide a floating nursery for walrus pups as the adults search the seabed for clams in the coastal waters. In August 2009 the Center for Biological Diversity said they have enough documented scientific proof that walruses are suffering from the negative impacts of Climate Change. The environmental group filed a motion with the Fish and Wildlife Service to give the Pacific walrus protection under the Endangered Species Act. In 2008 the polar bear, which also relies on sea ice, was listed as a Threatened Species.

The Pacific walrus has a population estimated around 200,000, double that in 1950. The Atlantic walrus, a subspecies found in Canada, Norway, Russia and Greenland, only numbers around 22,000. The Atlantic walrus never recovered from industrial hunting.

Offsets and Forestry Management

Forest preservation in U.S. South and Tropical Rain Forests to create Carbon offsets. Two separate initiatives to create Carbon offsets from forest preservation were announced in 2009. The Dogwood Alliance, a group of environmentalists, has joined forces with former foes like Staples and Home Depot to create Carbon Canopy. A report was released by a blue ribbon government commission encouraging U.S. corporate investment in tropical rain forest preservation.

The new alliance awards Carbon offset credits to Southern tree farmers who follow the Forest Councils Standards. Southern U.S. forests are 90 percent privately owned. The South is the largest paper and wood producing region in the world. In the past tree farmers would clear cut the timber and replant with fast growing pine, less environmentally friendly than the native canopy. The Carbon Canopy group awards credits to those who do not clear cut and replant with pine and only selectively cut.

The U.S. Commission on Climate Change and Tropical Forests released a report saying U.S. corporations investing $9 billion by 2020 to preserve tropical rain forests would generate $50 billion in savings for the firms by Carbon reduction avoidance and needs for Carbon Controls on their operations in the U.S. Forest offsets are criticized by environmentalists calling them a corporate accounting gimmick that lets the companies continue to pollute with no Carbon Controls on their operations in the U.S. and around the globe.

The report seeks $1 billion investment over the next three years in tropical rain forest preservation. The report also urges the international Climate talks to include Tropical Rain Forest Carbon production and avoidance from clear cutting and burning. The combustion and burning of rain forests makes up 20 percent of global Carbon emissions.

New England dairy cows return to natural grasses to reduce Methane. Local dairy farm owners, most notably organic milk producer, Stonybrook Farms of Vermont, have returned to feeding their cattle on natural grasses life alfalfa and flaxseed instead of corn and soy. The natural grasses replicate the spring grasses the cows evolved long ago to digest. Within six months at a farm in Vermont, Methane emissions had declined by 18 percent.

Cows have digestive bacteria in their stomachs that cause them to belch Methane. Methane is the second most powerful Greenhouse Gas behind Carbon Dioxide in Global Climate Change. Cows are tested in an air-tight tent enclosure to measure their "eruptions". Cows through mostly burps but some flatulence emit between 200 and 400 pounds of Methane a year.

World wide production of milk and beef is expected to double in the next 30 years. The United Nations has named livestock as one of the most serious near-term threats in Global Climate Change. When including the effects of deforestation to create pasture land, it is estimated that cows may be more dangerous to the atmosphere than cars and trucks combined.

Dairy Management, Inc., the research and marketing arm of the American dairy industry, says the American dairy industry accounts for two percent of Greenhouse Gasses from cows. It has a program called "Cow of the Future" looking at ways to reduce cow Greenhouse Gas emissions by 25 percent by 2020. Everything from cattle genetics to cow stomach bacteria is studied. While jokes about cow belching and flatulence will go over good with the kids, the problem of cattle Greenhouse Gas emissions is more serious than thought 20 years ago.

Point Carbon reported enough clean Carbon credits through world wide forestry management through 2020. The U.S. Energy Secretary Steven Chu reports enough clean Carbon credits through Energy Conservation alone to meet the 2020 17 percent Carbon reduction goal alone.

This means we need no increased consumer costs until at least 2030.

This means only workers just entering the work force will pay for new technology like clean coal, nuclear, wind, solar, biomass, and tidal, hydro and smart grid.

So why is U.S. industry lead by the U.S. Chamber of Commerce running scared? Why are the Republicans not behind something that could energize a new generation of Republicans? Carbon Control enjoys wide support among all demographic ranges in the younger than 35 group, male, female, all ethnicities and races.

The Kerry-Lieberman Carbon Control bill coupled with the Cantwell-Collins bill, both bi-partisan efforts, would have given back Energy rebates directly to consumers through 2025 from Carbon credit sales.

In 2010 Greenpeace issued a report critical of Carbon offsets from forest preservation. In the late 1990's a group of environmentalists and polluters started the first ever widespread experiment to control Climate Change through forest preservation. The Noel Kempf of Mercado Climate Action Project did save the biologically plentiful region of 6,000 square miles from logging. Yet it has failed to reach its Carbon reduction target.

Greenpeace released a study that raised doubts about forest conservation in foreign countries to offset U.S. domestic Carbon pollution. The Greenpeace report said Noel Kempf's target was 55 million metric tons of Carbon pollution prevention across 30 years. Greenpeace says after studying the 10 plus years of data on Noel Kempf of the likely estimate of Carbon pollution prevention is 5.8 million tons.

The Greenpeace report notes the three corporate sponsors; American Electric Power, BP America and PacifiCorp overestimated the environmental benefit of the project.

Meanwhile, Norway pledged $1 billion from 2010 through 2015 to preserve Brazilian tropical rain forests. The U.S. House passed Climate Change control Waxman-Markey bill set aside 5 percent of money from the sale of Carbon allowance credits to preserve forests in other nations.

The Senate Boxer-Kerry Climate Change Control bill also included set asides for foreign forest conservation.

PacifiCorp's Kyle Davis, Director of Environmental Policy and Strategy, said the bills on the Hill will not make it possible for industries to obtain the two billion tons of offsets they require.

Ned Helmet, President of the Clean Air Policy Center, says the U.S. and international governments must work out the way forth to conserve tropical rain forests as a key to U.S. and international Climate Change Control programs.

Chapter 10: Economics of Climate Change Control

The price of Carbon Control for you and its price on the open market is discussed. EPA originally estimated a ton of Carbon equivalent emissions would cost between $8 and $11 in 2030 raising your monthly electric bill by $80 to $140 a year. Other reports say a mid-range 2050 Carbon equivalent ton would cost $20.

The Regional Greenhouse Gas Infinitive (RGGI) has had Carbon Dioxide auctions since January 2009. The RGGI is voluntary Carbon Dioxide, Climate Change Control program developed over five years that began operations in January 2009. Carbon equivalent tons are trading between $3 and $3.50 per ton. The RGGI is made up of states in the Northeast from Maryland to Maine, with Pennsylvania observing. The auctions have raised over $720 million directly pumped back into state Energy Efficiency programs, Green Buildings, conservation and similar programs. The RGGI has auctioned off 49 million tons of Carbon allowances or credits.

The Carbon secondary market is falling dramatically as a national program takes shape in Congress and at EPA. Carbon trading volumes are modest but have grown from 300,000 in December 2008 to 980,000 in March 2009.

26 companies hold Futures and Options contracts on the secondary market exchange, the Chicago Climate Exchange and the New York Mercantile Exchange.

Carbon reached a high on December 1, 2008 at $4.23 per ton. In March 2009 Carbon traded around $3.51 per ton. In April it fell to $2 per ton. On the Chicago Climate Exchange Carbon was trading for $1.20 per ton in early May 2009.

Why is the price of Carbon falling?

Most of the RGGI states are one million tons below the cap where industry would need to make Carbon Dioxide reductions. This is due to the Great Recession. Economic activity and production and consumption are at a low.

Also, both the national Carbon Dioxide market based Cap and Trade programs developed in Congress and under current law by EPA have undermined the cost of secondary credits. In order to pass the Climate Change Control bill, Congress gave away 80 percent of the Carbon Allowances to industry for free. The funds raised in the Northeast at a disproportionate regional economic impact are gone essentially as a Carbon Tax. Only $90 million would have been raised in a national plan. Industry over paid by $170 million in the Northeast.

Will industry gets its money back?

No.

Many environmentalists are against Cap and Trade. They argued that a pure, straight tax on Carbon with no trading is preferable. Others argue that there is no way to verify international trades without a Cap and Trade system. Tracking tax dollars in China would be like a Ponzi scheme investigation.

The government will only auction off 20 percent of the Carbon credits on the open market. This makes the price of Carbon relatively cheap.

The Sulfur Dioxide credits were given away in 1993. The program is very effective at reducing Acid Rain and Sulfur Dioxide emissions. The Nitrogen Oxides program is only a few years old and not fully functioning yet.

It appears though that the main goal of reducing Carbon Dioxide by 17 percent by 2020 is still in place even though some Democrats wanted it lowered to six percent by 2020. The RGGI has a target of 10 percent reduction by 2018. The Federal program will create more demand as the economy recovers and 2020 nears.

Without giving the majority of credits away, it is unlikely manufacturing states in the Midwest and coal producing states like West Virginia and Kentucky and petroleum chemical states of Texas and Louisiana would vote for a Climate Change Bill. The immediate cost to industry is severely lessened by giving most of the Carbon Credits away for free in the early years of the program.

What does this mean to your electric, natural gas, and oil bill? What does it mean to heavy industry? Let us just stick with average industry cost in the early compliance period (2010-2014) of $3.00 per ton of Carbon Dioxide equivalent emissions.

According to the Executive of Office of Energy and Environmental Affairs (EEA) in Massachusetts, the average New England home emits Carbon at a rate of 1,237 pounds a month for electricity, 359 pounds per month for natural gas, and 891 pounds for oil per month. So for rounding purposes, let's say the average New England home creates 1 ton of Carbon dioxide a month. That's $3 for your Energy bill, not $20.

How much for heavy industry? It is harder to do a commercial sector Energy usage breakdown. Alcoa, a major aluminum producer in the U.S., is behind the Climate Change bill as are other keys industries.

In 2007, Alcoa created 23 million tons of Greenhouse Gases. If EPA pegs the initial price at $20 per ton, that's a new operational cost of $460 million to Alcoa. If the cost of a ton of Carbon is $3 per ton Alcoa's cost is $69 million.
Alcoa believes the U.S. can lead the world in light, sturdy aluminum and clean coal technology, making up the Carbon cost in increased exports while retaining and growing valuable heavy duty manufacturing jobs.

We've been trading Sulfur Dioxide for almost 20 years and Nitrogen Oxides for several years. Have you noticed sharp inflation? Have you seen sharp non-gasoline Energy bills? No. And you won't under a market based Cap and Trade program.

The myth that a Carbon Cap and Trade program will damage the economy is an old industry line they have been using since the 1970's.

The truth of the matter is, Clean Air Act legislation is the most successful environmental program ever. It cost far less than the EPA and industry estimated. It created more economic growth and jobs than lost.

Whether you believe Climate Change is real or not, you do not have to fear big government placing a heavy tax on your home Energy usage or seeing consumer prices spike.

There were defections over Climate Change at U.S. Chamber of Commerce in 2009 and 2010. Apple was one of the notable corporation's to quit the U.S. Chamber of Commerce over the Chamber's opposition to Climate Change Control. "We strongly object to the Chamber's recent comments opposing the EPA's effort to limit Greenhouse Gases," wrote Catherine Novelli, Apple's Vice President of Worldwide Government Affairs in a letter to the Chamber.

Nike resigned from the Chamber over the issue of Climate Change Control and how to approach the problem. "We fundamentally disagree with the U.S. Chamber of Commerce on the issue of Climate Change," the company said in a statement, saying it was quitting the Chamber. "It is important that U.S. companies be represented by a strong an effective Chamber that reflects the interests of all its members on multiple issues," the statement said. "We believe that on the issue of Climate Change, the Chamber has not represented the diversity of perspective held by the Board of Directors."

Recent Chamber defections include Pacific Gas and Electric, PNM Resources, Exelon and Duke Energy.

The wrangling was enough to warrant response from the Chief Executive of the Chamber, Thomas Donohue, who said the industry lobby "continues to support strong Federal legislation and a binding international agreement to limit Carbon." Donohue said, "Some in the environmental movement claim that, because of our opposition to a specific bill or approach, we must be opposed to all efforts to reduce Greenhouse Gases or that we deny the existence of the problem. They are dead wrong."

The U.S. Chamber of Commerce is threatening to sue EPA about its position in controlling Carbon. "Our position is simple," the Chamber's says on its website in the "5 Positions on Energy and the Environment. There should be a comprehensive legislative solution that does not harm the economy, recognizes that the problem is international in scope and aggressively promotes new technology and efficiency.

Protecting our economy and the environment for future generations are mutually achievable goals."

The White House released a study compiling the current effects of Climate Change in the U.S. Among some of the effects are heavier down pours in the Midwest and East, more intense heat waves in the Northeast, changing migration patterns of butterflies in the West, and melting snow packs. The effects are expected to worsen in coming years, affecting coastlines, forests, farms, floodplains, water and Energy supplies and transportation.
The climate of New Hampshire will resemble North Carolina's and Illinois' that of Texas.

The report was compiled by 13 Federal Agencies and the White House. Under a 1990 law the report is required every 10 years. The impacts predicted are not much different than those reported in the 2000 study. Cited were more powerful tropical storms and coastal erosion caused by the melting ice caps. Also cited were more droughts in the Southwest. Smaller mountain snow packs effects water supplies across the West and Northwest, hydroelectric generation and impacts on fish spawning.

As Congress was trying to enact a Climate Change bill in 2009 and 2010 China was still seen as being unwilling to make Carbon reductions.

I worked on the Clean Air Act of 1990. I was questioned extensively by elected and appointed officials in several states, why should we act when there is no scientific consensus?

Some how, we acted. Cancer rates of all types are down two percent per year since 1990. Was cleaner, healthier air the cause? Not likely, but maybe a small contributor. Mostly scientific medical genius and life style changes made the cancer rate change.

So what is the public to think? What are politicians supposed to do? It is not clear. Wait? Alarmists would say species extinctions. Skeptics would say irreparable economic damage. Is there a middle ground?

Is Cap and Trade a tax? Well, no, Cap and Trade is not a tax. It is an indirect tax that will see your Energy bill rise $8 to $20 in the next two decades. Why an indirect tax? President Clinton tried to enact a $0.25 per ton straight Carbon tax in the early 1990's and it failed miserably in Congress.

Since there is no known effective, widespread, in-practice Carbon Control technology like the Sulfur and Nitrogen markets have, you will pay for Carbon Control, either way.

The only tried Carbon Control technology is Carbon sequestration, or injecting Carbon dioxide into under ground wells or caves. The coal industry has had 20 year to develop this technology and no progress. The coal industry is relying on small test pilot programs for Carbon injection underground from Stimulus funds.

The Congressional Budget Office presented a shoddy, lopsided cost assessment of House Climate Change Bill with no assessment of benefits.

In a shocking disgrace to professional, neutral economic analysis, the Congressional Budget Office presented a cost impact assessment of the House Waxman-Markey Climate Change Control bill while negligently ignoring the health, property, environmental, geographic and job creation benefits of the bill.

Douglas Elmendorf, Director of the Congressional Budget Office said economic growth will decline modestly over the next few decades and cause "significant" job losses. He conceded the analysis did not include benefits of stopping Climate Change.

EPA would be hammered for presenting a benefit only assessment of any measure while neglecting the cost impact.

The University of Massachusetts Amherst in Senator Markey's home state says the forthcoming sustained growth of Green Technology jobs and economic growth in the American Clean Energy Security Act (ACESA) will create 1.7 million new jobs.

Hearings and an investigation should be launched for fraud at the Congressional Budget Office and Mr. Elmendorf should be forced to resign for his biased work.

Economics of Climate Change control was debated in the Senate in 2009 and 2010, not the science or validity of Climate Change.

Fence sitting moderates of both parties were important to passage of Climate Change Control legislation in Congress.

Senator Lindsey Graham (R-SC), who co-wrote an Op-Ed piece in the New York Times supporting Climate Change Control legislation with Senator John Kerry (D-MA), said the U.S. needs an environmental policy on Climate Change that brings better economic policy.

There are conflicting estimates of the costs of Climate Change Control legislation and no one has reported on benefits. EPA puts the cost at $0.30 per day per household. The Congressional Budget Office Director Douglas Elmendorf said October 14, 2009 that the House passed Climate Change Control bill would create "significant" job losses in the fossil fuel industries. Elmendorf said the House bill would stall economic growth between 0.25 percent and 0.75 percent per year through 2020 and harm the economy for decades.

The Congressional Budget Office report on the House bill did not account for benefits of the House Climate Change Control bill, like new Green Technologies, domestic job growth, national security, internalizing the external societal costs and harms of free Carbon pollution, health and property loss damages avoided and societal benefits.

The negative impacts reported by industry groups are questionable.

The American Council for Capital Formation and the National Association of Manufacturers say that 2.4 million jobs will be lost, assuming only half as many Carbon offsets in the future than that reported by EPA that would keep Energy costs low.

A report by Charles River Associates for the National Black Chamber of Commerce forecasts 2.2 million job losses by 2030 due to the higher costs of nuclear and geothermal Energy projects.

What is needed now is full projection of the economic, security, social, cultural and environmental benefits of Climate Change Control legislation.

EPA says Greenhouse Gases are a public health threat. In early December 2009 as the Copenhagen Protocol Conference kicked off in Denmark the U.S. Environmental Protection Agency ruled that Greenhouse Gases like Carbon Dioxide and Methane are indeed public health hazards and warrant EPA regulation.

The Supreme Court ruled under President Bush that Greenhouse Gases (GHGs) are governable under the Clean Air Act, and instructed EPA to do a public health threat assessment of GHGs. The move sends a strong signal to the world community that the United States will regulate GHGs soon.

This was a strong signal to Republicans to act swiftly in 2010 to pass a Climate Change Cap and Trade Carbon bill. Most analysts agree a Carbon Cap and Trade program would be cheaper for industry to comply with especially given the excessively generous give always in the House bill from June 2009 than an Operating Permit program run by EPA.

Operating Permits for Criteria Pollutants like Ozone, Particulate Matter and Nitrogen Oxides have been around for decades and work extremely well when enforced.

Sulfur Dioxide has had an emissions trading program since 1993. Nitrogen Oxides also have a trading program called CAIR, Clean Air Interstate Regulation. These two commodities trade for about $700 a ton on the open market, far below the $2,000 a ton mark estimated by industry and the $1,500 per ton by EPA. By contrast Carbon is expected to trade at an extremely cheap $15 per ton.

EPA will move to impose the Best Available Control Technology (BACT) on GHGs, a time proven cost effective control strategy for traditional pollutants. But EPA could be sued by environmentalists to impose the Lowest Achievable Emissions Rate (LAER) standard since GHGs pose such a wide spread both health and property damage problem.

This is highly likely. This is the closest the Federal government has come to out right banning coal in its history. Dirty coal is both a huge GHG problem and also a serious public health threat that presents a clear and present danger to adjacent residents of coal fired power plants both to air they breathe and the water they drink.

For the time being in early 2010 EPA said it was willing to wait and work with Congress on GHG regulation. Obama tried to use EPA regulation to force the Republicans in Congress to promulgate Climate Change Control legislation.

EPA will regulate Greenhouse Gas emissions in January 2011. The Agency covered Greenhouse Gases under existing Criteria Pollutant Clean Air Act regulation for stationary sources.

Greenhouse Gases are covered by New Source Review and Prevention of Significant Deterioration. Both rules have withstood several Court decisions successfully.

In January 2011, EPA will regulate all six Greenhouse Gases in major sources creating more than 200 tons (1 ton is 2,000 pounds) per year. Two years later in 2013 EPA will cover Greenhouse Gas sources creating more than 100 tons per year of Carbon equivalent pollution. Two years later in 2015 EPA will cover virtually every sector of the economy generating more than 25 tons per year of Carbon equivalent Greenhouse Gases.

The Obama Plan carefully introduces the new coverage in non-election years. EPA Administrator Lisa Jackson said in a press release in early 2010 that the Agency will not regulate Greenhouse Gases before January 2011. Moderate and conservative Senators of both parties are backing away from stripping EPA of regulatory power over Greenhouse Gases after careful legal review.

The House Climate Change Control bill is an industry boondoggle, giving nearly all pollution licenses away for free. The Senate Kerry Climate Change Control bill, introduced last year, mirroring the House Bill, is dead in the Senate.

The Obama Plan mandates selling pollution licenses to industry and then letting the over the counter trade exchanges of NYMEX and the Chicago Climate Index set the value of Carbon.

Most Carbon traders agree Carbon will trade for around $12 to $15 a ton through 2020, eventually stabilizing around $20 a ton in 2030. Traders expect volatility and bubbles in the Carbon market.

As offsets from reforestation and forestry management in timber rich Rain Forests in Brazil and Indonesia and Energy Conservation and Efficiency credits evaporate around 2020, the real bite for industry and the consumer begins.

In 2020 major new source of clean coal, nuclear, major wind, solar, tidal, and hydro and biomass projects along with cleaner petroleum and natural gas, will come onto the upgraded grid and begin to displace imported transportation oil and reduce Carbon towards the 2050 80 percent goal below 2005.

The Obama Plan estimates the average annual utility bill will increase by $8 to $10 a month. A modest Carbon tax on motor fuel will induce alternatives and more mileage cars.

The Obama Plan rebates the Carbon license fees directly to consumers, zeroing out any new cost for Carbon Control to the average home. Some of the fee is invested in clean technology.

The Obama Energy Plan had bi-partisan endorsement.

A panel of international experts on Carbon trading gathered in New York City in October 2009 to discuss and outline what a U.S. Carbon trading market might look like. The first half of the debate focused on how legislation, policy and regulation will affect the Carbon market. The second half of the debate centered on the role of electronic trading in the Carbon market.

The panel debate was sponsored by Trayport, a London software company specializing in Carbon trading platforms. Trayport software is used by 70 percent of the Carbon trading market worldwide.

Several themes emerged during the discussion, the role of Carbon offsets and how to seamlessly create them, the expected price of Carbon per ton on the open market, the need for transparency in operating markets and trading, regulatory consistency and neutrality on Carbon market price and the stages of growth in Carbon trading in the U.S. and world.

Carbon offsets come from new players to the Climate Change Control debate, tree farmers, foresters and cattle ranchers. Offsets are a cost effective way to reduce Carbon in the near term through 2020. Many agreed available funding would not meet required market investment needs to create the offsets worldwide.

Environmentalists are at odds over Carbon offsets produced both domestically and internationally. Greenpeace released a report critical of Carbon offsets through forest preservation from a 10 year old experiment in Brazil called the Noel Kempf Climate Action Project.

Christopher Hunter, Vice President at Climate Change Capital says, "Increase domestic offset and Carbon reduction programs. Don't import them."

The panel also agreed the price floor in the Waxman-Markey bill would allow the back log of offset projects to receive funding and get started while stabilizing the price of Carbon on the open market. The panel agreed the price floor for Carbon is between $15 and $20 per ton through 2020, thus greatly blunting the cost impact on the U.S. economy and consumers.

Several lessons from the European Union trading experience were listed. First and foremost is the need for transparency in market function and trading. The European Union gave away all Carbon allowances in Phase 1, but several companies did not pass along the savings to customers and thus reaped vast profits, angering the public. As the European Union moves towards Phase 3 it will auction off all allowances, a market mechanism pioneered by the northeast Regional Greenhouse Gas Initiative in January 2009.

Professor Praveen Kumar of the Bauer College University of Houston says, "The uncertainty is not in model price but in regulation. There will be price bubbles and crashes in the Carbon market. If bureaucrats "print" more free allowances it could destroy the long term Carbon market."

The panel generally agreed there will be no accord passed in the U.S. The earliest more international Carbon markets could open is 2013.

Between now and 2020 the panel sees industries, governments and traders going through a learning phase and gaining lessons from mistakes.

The outlook for Carbon trading in the U.S. looks bright as long as the market and regulators can manage the growing pains.

President Barack Obama announces $2.3 billion in Clean Energy Manufacturing tax credits. President Obama announced a new initiative under the Stimulus bill called the Clean Energy Manufacturing Initiative. The President said the push promotes good paying, sustainable, domestic jobs while creating Energy independence and improving national security.

Obama noted how China has made a commitment to Clean Energy, becoming the world's largest government investor in Clean Energy. Obama also said Germany leads world market share in solar panel production. Virtually all of the batteries we use from everything from laptops to hybrids and electrics are manufactured in Japan or Asia. The President's Clean Energy Manufacturing Initiative aims to help the U.S. economy catch up with Europe and Asia and bridge the gap in public and private Clean Energy investment in the U.S.

The $2.3 billion in tax credits will go to wind turbine, solar panel, and advanced electric battery manufacturing plants creating tens of thousands of jobs, stimulating new manufacturing plant construction, and the rehabilitation and modernizing of older industrial facilities that can be reused for Clean Energy manufacturing.

180 projects from 40 states will receive the tax credits. One example, TPI Composites in Newton, Iowa will expand, hiring 200 new workers and building a new plant in Nebraska. The $2.3 billion government investment will spur a matching $5 billion private sector investment, creating more than 17,000 new jobs in the Clean Energy manufacturing sector, at a time when unemployment remains stuck at 10 percent and Americans are clamoring to work.

President Obama said the government received $7.6 billion in grant requests but was only able to fund the $2.3 billion. Obama is requesting an additional $5 billion from Congress to meet the pent up demand of the Clean Energy manufacturing sector.

The winners and losers under a Carbon Cap and Trade law are forecast by a consulting firm. Point Carbon, a market analyst firm, identified in a report released today the winners and losers of the Senate Carbon Cap and Trade bill. The power and oil sector represents 40 percent of covered Carbon emissions. Southern Company is the biggest loser and Excelon Corporation the biggest winner.

The power sector is most vulnerable. Southern company would pay 12 percent of its operating income to comply with Cap and Trade, American Electric Power would pay 11 percent and Duke Energy would pay five percent. American Electric Power would pay $2.3 billion annually and Southern company $2.2 billion.

ExxonMobil is the largest Carbon emitter in the U.S. at six percent of total national Carbon emissions, but its operating income will only slightly suffer under a cap and trade law.

Based on the price of a ton of Carbon projected at $15 per ton, ExxonMobil would pay $5.9 billion annually to comply. Raising gasoline prices five percent would leave ExxonMobil's total cost of compliance at $277 million per year out of its $84.1 billion operating income. A $15 per ton Carbon price translates to an increase of $0.13 per gallon of gasoline.

Some power companies gain under a Cap and Trade system. Excelon would see a net increase of 36 percent in its operating incoming due to its large portfolio of low emission power generation plants. Other Energy company winners are FirstEnergy and PG&E.

What should the government do with Carbon Allowances? An advisory committee to the California Air Resources Board recommended the state provide cash rebates to residents from the proceeds of Carbon credit sales.

The Regional Greenhouse Gas Initiative (RGGI) from Maryland to Maine in the Northeast auctions Carbon Allowances. The RGGI uses the millions of dollars to fund state Energy conservation, renewable Energy projects, and Green Technology projects like solar and wind. The RGGI sells credits quarterly. The price has hovered around $3 per ton plus or minus a dollar since it began operating in 2009.

In Europe where a Carbon trading program has been operating for several years the price per ton of Carbon fell to below $15 per ton from $20 after the Copenhagen Protocol Collapse in mid December, 2009. Europe and RGGI are the only two functioning Carbon markets in the world. Europe has an active trading program but is still fine tuning it.

The RGGI and California set their emissions control baseline as 2005. Carbon must be lowered by 10 percent to 2005 levels by 2018. The targets will likely be missed. The targets are highly laudable and ambitious. Real Carbon reductions will occur by 2020.

The Western Climate Change Initiative (WCI) composed of several Mountain and Western states and most of the Canadian provinces stretching as far as the East Coast is still in the pre-deployment stage and presumably will follow California's lead.

There is another voluntary state led regional Carbon trading consortium in the industrial and coal heavy Midwest. The U.S. voluntary Carbon trading and reduction consortiums and Europe comprise about 30 percent to 40 percent of global Carbon Dioxide emissions yearly.

GFI Group, a major Wall Street brokerage house and the leading Carbon trader on the secondary European market expects California will have the first active secondary Carbon trading market in the U.S. by 2012. There is no secondary commodities market trading of the RGGI credits because of the Great Recession of the last two years. The Carbon Cap limits have not been reached so no one needs to buy offset credits.

It is unknown if the California market will function on the secondary exchanges like the New York Mercantile Board and the Chicago Climate Exchange. Carbon was trading around $2 per tons on the open secondary market before the House Waxman Markey Climate Change Control Bill passed the House in June of 2009. The bill gave away 75 percent of Carbon credits for free to industrial polluters. In 2010 Carbon trades for pennies a ton on the commodities boards.

EPA is wrapping up a multi-year effort to write a rule covering Carbon equivalent Greenhouse Gases (GHG). A court challenge when the rule is final is certain. The Supreme Court directed EPA to regulate Carbon equivalent GHGs under the Clean Air Act. EPA, often pre-empted by Congress and Court decisions, proceeded with its normal rule making process.

The rule covers two-thirds of Greenhouse Gases at major oil refineries and utilities. About 1,000 new permits are required by industry. EPA is using its Court tested Best Available Control Technology or BACT to regulate Carbon.

BACT is an econometric engineering analysis. BACT considers commercial off the shelf control technology, cost, benefit and other impacts. There are currently no market ready Carbon Control technologies at any price.

The rule comes out of the stationary source group created by President Nixon. The rule does not create a Carbon Cap and Trade emission permit system.

Congress debated such a scheme in 2009 and 2010. The BACT Carbon Control rule covers new plant construction creating 100,000 tons per year of Carbon equivalent of Greenhouse Gases. The usual operating permit for Criteria Pollutants is 250 TPY. A cost-benefit analysis by EPA staff determined that a 250 TPY Carbon equivalent threshold was economically unfeasible since it would cover restaurants and apartment buildings. Plants expanding operations adding 75,000 tons per year or more would need a permit, regardless of size.

Chapter 11: The Regional State Voluntary Climate Control Initiatives

So what is going on at the state and regional level in the United States for Carbon Control? A lot more is happening at the state and province level than you realize. Currently 23 states are participating or are planning on a joining a voluntary regional Climate Change Control group and a dozen others are watching intently. These states and most of Canada and northern Mexico are voluntarily binding together in regional agreements to cut Carbon pollution.

In the next 10 years this will represent at least half to two thirds of the entire United States GDP and economy.

This is a major development and far more stringent than what Congress is considering. The United States Environmental Protection Agency is incrementally and slowly taking the same approach nationally that more than half the states are already engaged in performing. The EPA through years of rule making and political posturing on the national stage will capture the South which remains largely outside of the voluntary agreements so far.

California represents 10 percent of the United States entire population. The Northeast from Maryland to Maine represents another 30 percent of the United States population. The Regional Greenhouse Gas Initiative operates in this region.

The RGGI has a Carbon equivalent cap of 188 million tons between 2009 and 2014 decreasing by 2.5 percent per year through 2018 for a 10 percent reduction over 2014 by 2018. The RGGI Carbon Cap and Trade system currently only covers power plants of the largest electric producers regardless of fuel source. The RGGI auctions Carbon allowances or credits at around $3.00 per ton.

The RGGI began operating in January 2009. Its first compliance deadline for the large power plants is December 31, 2011. The RGGI has raised $730 million dollars in revenue from the sale of Carbon Allowances or assets on the open market.

Carbon can be bought on the Chicago Climate Futures Exchange for about $0.05 to $0.10 per ton. Many analysts project the price of Carbon will rise to $15 by 2020 and to $30 by 2030.

By contrast Sulfur Dioxide traded since 1994 sells for about $750 per ton. This first Cap and Trade program is successful in reducing Acid Rain that kills fish, destroys forests, damages stone, steel and cement, and poisons public water supplies.

Nitrogen Oxides also trade around the same price as Sulfur but have not been trading as long. Nitrogen Oxides also cause Acid Rain and Ground Level Ozone or Smog that hurts the delicate lung on warm summer days.

The RGGI states must use 25 percent of auction sale proceeds to benefit consumers.
66 percent of RGGI auction sales proceeds are used by member to states to sponsor local Energy Efficiency programs and Renewable Energy projects.

Several states have found that for every $1 invested in Energy Efficiency a multiplier of $4 to $6 is created in Energy savings and new job growth.

The United States Department of Energy reports that for every $1 million invested in Energy Efficiency and Renewable Energy 50 construction jobs are created and an additional 20 design and manufacturing jobs are created.

The RGGI Carbon Cap and Trade program has created 51,100 new jobs in its first two years. It holds its Carbon Allowance auctions quarterly. The Western Climate Change Initiative and the Midwestern Climate Change Accord are modeling themselves after the RGGI as is the European Union Emissions Trading System (EU_ETS) in its latest phase.

The United States Environmental Protection Agency will delegate its monitoring and enforcement role to member states through these voluntary regional agreements. The associations themselves are merely administrative coordinating bodies for member states and do not provide technical assistance. Much of the trading activity and auditing activity is carried out by contracting companies in Massachusetts and Virginia.

The EPA traditionally delegates much of its authority to state agencies to run air quality control programs. The EPA's main role is regulatory development and harmonization between different regions of the national economy and technical support.

EPA still plays a vital role in permitting, monitoring and enforcement through its 10 regional offices around the country. The regional offices coordinate the efforts of the various states in planning, permitting and enforcement.

Greenhouse Gases and Climate Change Control is no different than any other air pollutant mandated by Congress or the Supreme Court. The Conservative Roberts Supreme Court ruled in July 2007 that Carbon equivalent gases are subject to the Clean Air Act.

Climate Change Control is here today. Programs are up or soon will be from Maine to Iowa to Nebraska to California. The RGGI has paved the way.

EPA will rein in Texas and the Petroleum Coal Belt from Virginia to Texas. Carbon Control is here and it is working. The cost is 1 percent of utility expenses. The cost to you today is $0.73 on you electric bill in the Northeast.

Chapter 12: The Politics of Climate Change Control in 2009 and 2010

Why is Environmental Protection Political?

The standard theory is that Democrats are pro-environmental protection and that Republicans are against it in favor of business interests. Is this true or just a hackneyed argument meant to drive a wedge between all Americans?

Why is the environment so easily a politically divisive issue? Environmental issues are incredibly complex, involving science, human health, property, real dollars and often jobs.

Environmental issues first came to the public attention in the late 1950's and early 1960's as our industrial lifestyle began having larger impacts on human health in the growing United States population.

As the 1960's unfolded the environmental issues of the day were picked up by the Protest Movement and seen as a rallying cry for all of the various social movements of the 1960's. Environmental problems were easily laid at industry door.

Environmental pollution was seen as technically capable of solving with some minor modifications to operations and behavior.

The country was untied in protecting human health, wildlife, fisheries, habitat and private property.

President Richard Nixon embraced the environmental and safety protection issues of the Protest Movement. President Nixon established dozens of Federal and State regulatory agencies in environmental and safety protection.

As the recession plagued 1970's wore on the Democrat controlled Congress passed numerous Federal environmental laws. They covered every type of human health problem from industrial pollution.

Safety became increasingly more popular and easier to sell to the public than environmental protection as the decades unfolded.

In the 1990's the Republicans took control of Congress and simply tabled most new or updated environmental regulation.

The success of many environmental programs and medical advances lead to cancer rates declining. Cancers are more survivable. Fewer Americans smoked tobacco and became more aware of steps they could take to improve their own health.

After nearly 20 years of environmental law abeyance, cancer rates among non-smokers under the age of 35 are increasing again.

Is this due to outdated environmental law and new technologies poisoning us in ways that we do no know or are not willing to regulate?

Safety is seen as worth paying for because it protects human health and property in the day to day present world. The environment and the benefits of regulating industry today have a very steep immediate economic cost with diffuse and unknown future reduced cancer rates and lessened health impacts.

This is a price that the politicians think is not worth paying. Politicians vote on things that provide immediate benefits and costs later on down the road. This is the road map of lobbyists and successful legislative campaigns. Paying now for benefits in the future flies in the face of political science especially in the short election cycles of the United States.

The environment is easily dismissed as an emotional issue with hard immediate economic impacts, job losses and closed industries. While this is rarely the case, it is an easy sound byte that gets viewer and voter attention. Whenever an environmentalist tries to make a sound byte or even a full length film on Climate Change like Al Gore, he is accused of using scare tactics and fear mongering.

Can Energy Policy Change Carbon Output?

After 37 years of failed Energy Policy the U.S. should pass a Climate Change Control bill. Where are we today on petroleum independence?

We have made no progress. Okay. So you don't believe Climate Change is real. Is imported petroleum a myth too?

In almost 40 years of promoting Alternative and Renewable Energy the U.S. has made zero progress. Carbon fuel still makes up 99 percent of our Energy use, forgetting nuclear. How do we make inroads on imported petroleum dependence to increase our economy, mobility, safety, security and environment?

The Stimulus bill is doling out millions of dollars to states like Massachusetts to build Renewable wind turbines and solar facilities. It will take decades more of funding like this to make a dent in petroleum's dominance of the U.S. market.

Is the Cap and Trade bill a step in the right direction? Yes. It is an indirect tax on Carbon fuels that provides more incentive for utilities, Big Oil and Big Auto to invest in non-Carbon Renewable and Alternative, Green Technologies.

The House proposed giving away 85 percent of the Carbon credits for free. Under that plan the cost would not be passed on to consumers for years. No need to worry about worsening the Recession. After Stimulus funding for Green projects runs out where will Green Technology companies get their funding? After four decades, the entire industry is still in "start-up" mode.

The time to act is now, not only by passing the Carbon Cap and Trade bill, but by continuing sustained Federal investment in Alternative, non-Carbon technologies. Do you really still want to be at war in the Middle East in 2020? I don't think so.

10 moderate Senate Democrats opposed the Climate Change Control bill. The coal and manufacturing dependent states sent a letter to President Obama opposing the Climate Change legislation. The Senators said they may support a bill if it had protections for American industries against foreign countries like China, India and Brazil that are not planning on adopting mandatory Carbon limits. The Senators said they want to impose tariffs on countries without Carbon limits to protect American heavy industries. Obama opposes retaliatory Climate Change tariffs on countries failing to adopt Carbon limits.

The Senators were vital to passing a Senate version of the Waxman / Markey Energy bill passed in the House in June 2009. The Senators were from Midwestern coal producing states like Kentucky, West Virginia and Ohio.

The Senators included Evan Bayh of Indiana, Sherrod Brown of Ohio, the deceased Robert C. Byrd and John D. Rockefeller of West Virginia, Bob Casey and Arlen Specter of Pennsylvania, Russ Feingold of Wisconsin, Al Franken of Minnesota, and Carl Levin and Debbie Stebenow of Michigan.

They asked for transition financing for Energy intensive industries as rebates on Energy costs, strong international Carbon limits agreements, Carbon emissions monitoring programs of other countries and strong funding of Clean Energy.

States won Federal case to sue over Carbon. A Federal New York Appeals Court affirmed the rights of states to sue utilities over Carbon emissions. The case came down overturning a 2005 District Court decision that ruled the issue was political not judicial.

Meanwhile EPA plods along on its rulemaking process to control Carbon equivalent emissions. It is unclear how far EPA will go.

The Second Circuit U.S. Court of Appeals in New York found in favor of eight states - California, Connecticut, Iowa, New Jersey, New York, Rhode Island and Wisconsin - saying they could proceed with a lawsuit against American Electric Power, Southern Corporation, the Tennessee Valley Authority, Xcel Energy, and Cinergy, all big coal burning utilities.

First brought in 2004, the states called the utilities a "public nuisance" and sued to reduce Carbon emissions widely held to cause Climate Change.

The utilities said only Congress can decide if Carbon causes Climate Change. American Power Electric spokesperson said they had not decided whether they would appeal.

Pat Hemlepp the spokesperson said, "We don't feel litigation is an appropriate avenue to address Climate concerns. In our view, it's a policy issue. Legislation would be the best approach and that's happening now," Mr. Hemlepp said.

National Carbon changes from 1990 to 2010. Russia proposes to reduce Carbon by 10 to 15 percent by 2020 over 1990 levels. Soviet Carbon production fell by 40 percent in the 1990's and remains about 35 percent below 1990 now. They are set to grow again. Europe's Carbon production has remained fairly constant since 1990, due to heavy taxes on fuel and a switch to more natural gas usage.

Japan's Carbon production is up by 9 percent over 1990. The United States Carbon production is up 7 percent since 1990. Since Kyoto two decades ago global Carbon emissions are up 30 percent. Chinese Carbon emissions have doubled from 13 percent to 27 percent. The U.S. emits about 25 percent of global Carbon. China and the U.S. are responsible for 52 percent of all global Carbon emissions. The next largest producers of Carbon are India and Brazil. All three of these countries have made strong voluntary pledges in the last year to reduce Carbon, a major Obama victory. But China's Carbon output could increase by 40 percent by 2020.

The Maldives may sink because of Climate Change, but temperatures in northern climates will become more comfortable. It will create new shipping routes making trade easier, cheaper and will open the Arctic. This will boost the world economy with new mineral and oil deposits in the more accessible and plentiful Polar Regions. There will be winners and losers in Climate Change.

Global 350 is worldwide movement that seeks to return worldwide Carbon levels to 350 parts per million (ppm). We are now at 387 ppm.

EPA estimates between $1 and $18 billion in revenue from offsets sold in the Carbon cap and trade bill on the Hill now, even though its only 15 percent of the total Carbon credits to be allocated. The money will be rebated to consumers to offset increased utility bills and also fund Green Technology projects.

The European Union agreed to pay $100 billion annually by 2020 into a global fund to reduce Carbon in poor nations. The EU failed to make any public funding a part of this pledge upsetting environmentalists and UN officials.

New Zealand is the only non-EU nation to have a Carbon cap and trade program. The EU environmental commissioner has called on the EU to make a 30 percent Carbon cut over 1990 by 2020 in comparison to current 20 percent agreement. Britain, Denmark, Netherlands and Slovenia agree but big coal using countries Poland and Italy oppose it.

EPA's finalized its rule on New Source Review of Greenhouse Gases in 2010.

GHGs are covered by the National Ambient Air Quality Standards. The historical regulating threshold for Hydrocarbons and Nitrogen Oxides is 100 and 250 tons per day. The threshold for Greenhouse Gases like Carbon Dioxide and Methane and four other smaller contributing chemical GHGs under New Source Review and Title V of the Clean Air Act of 1990 is 25,000 tons per day, covering a vast array of industries. Operating Permits and New Source Review compliance costs for industry are far more costly than a Cap and Trade Climate Change Control Act.

The utilities know this. The oil, auto and coal industries know this. The EPA is following a Supreme Court order. Industry is petrified. Polluting industries did not have to comply with New Source Review and Operating Permits under President Bush. Now that EPA is enforcing these laws. Texas and the heavy industry states will see severe economic hardship and job loss. The cost of compliance with Federal law and the Clean Air Act that they ignored for the last decade comes to painful screeching halt.

A Market Based Incentive Program like cap and trade is the least cost compliance option available to industry. This is especially true since Congress planned on giving most the Carbon credits away for free.

EPA will sell all of the Carbon allowance permits as it did with Sulfur and Nitrogen Cap and Trade programs. These two Cap and Trade programs came in very cost effectively, achieved great environmental protection, health and property benefits.

The Republicans may try to delay things. Some industries may sue. But there will be Cap and Trade. It works. It is proven. It is the cheapest program. It will be accomplished sooner than the Skeptics think. Industry, particularly the insurance industry, will buy the lobbyists to get the law of Congress enacted.

www.ingramcontent.com/pod-product-compliance
Lightning Source LLC
Chambersburg PA
CBHW051446170526
45166CB00001B/133